# Cable engineering for local area networks

BARRY J ELLIOTT

WOODHEAD PUBLISHING LIMITED
Cambridge England

Published by Woodhead Publishing Ltd
Abington Hall, Abington
Cambridge CB1 6AH, England
www.woodhead-publishing.com

First published 2000

British Library Cataloguing in Publication Data
A catalogue record for this book is available from the British Library.

ISBN 1 85573 488 5

Cover design by The ColourStudio
Typeset by Best-set Typesetter Ltd., Hong Kong
Printed by St Edmundsbury Press, Suffolk, England

# Contents

# Preface

This book is presented as a formal textbook for students studying courses such as the NVQ (National Vocational Qualifications), City & Guilds and BTEC courses, and BSc courses in communications engineering in the UK, and the courses for Certified Electronics Technician, Certified Fiber Optics Installer, Certified Network Systems Technician and Telecommunications Electronics Technician in the US. It also serves as a background reader for the US BICSI® courses.

The book covers material that students would be expected to know for the Datacommunications courses of the American Electronics Technicians Association and the British City & Guilds course 3466, Copper and Optical Communications, plus most other telecommunications and datacommunications C&G courses such as Telecommunications and Electronics Engineering (2720, 2760 and 3478). The book is also aimed at NVQ (and SNVQ) students studying copper and fibre communications technology, levels one to five, and also for any future qualifications generated by the European Institute of Telecommunications Engineering and the European Intelligent Buildings Group. Students studying for BICSI qualifications, such as RCDD, will also find this book helpful, especially in the area of European and International standards.

Students studying BTEC and degree courses on electronic and communications engineering will, from time to time, require some link between the theoretical and the practical. Hopefully they will find it between these pages. Readers will certainly come across technical definitions of communications theory and practice never before gathered together within one book.

Apart from use as a formal textbook it is also presented as an aid for IT managers, consultants, cable installation engineers and system designers who need to understand the technology and physics behind the subject and the vast panoply of standards that accompany it. The book does not present itself as a design manual for structured cabling but rather explains the terminology and physics behind the standards, what the relevant standards are, how they fit together, and where to obtain them from. Anybody studying this book will be able to read the standards, understand manufacturers' data sheets and their conflicting claims and be suitably equipped to address those problems raised by the need to design, procure, install and correctly test a modern cabling system, using both copper and optical fibre cable technology.

*Credo ut intelligam*
*Barry J Elliott BSc,*
*MBA, RCDD, CEng, MIEE*

# Acknowledgements

The rights of all trademark holders are acknowledged. All diagrams are by the author (except where acknowledged otherwise) although many are based on original drawings by Ian Torr and Martyn Davies of Brand-Rex Ltd and my thanks go to Ian Mack, Managing Director of Brand-Rex, for permission to use them. Also to Les Carter of Brand-Rex, the 'main-man' of copper cables, for his inestimable knowledge, to Vince Mahoney for his suggestion that I could do it, and to Alan Flatman of LAN Technologies for the glimpses into the future of standards. Finally, thanks to Yvonne for making the space.

# List of abbreviations

| | |
|---|---|
| ACR | Attenuation to crosstalk ratio |
| ADC | Analogue to digital converter |
| ADSL | Asymmetric digital subscriber line |
| ANSI | American National Standards Institute |
| ATM | Asynchronous transfer mode |
| AWG | American wire gauge |
| BER | Bit error rate |
| BERT | Bit error rate tester |
| CATV | Community antenna television (cable TV) |
| DAC | Digital to analogue converter |
| dB | Decibel |
| EIA | Electronic industries alliance |
| ELFEXT | Equal level far end crosstalk |
| EMC | Electromagnetic compatibility |
| EMI | Electromagnetic immunity (or sometimes 'EM interference') |
| ESD | Electro static discharge |
| FCC | Federal Communications Commission |
| FEXT | Far end crosstalk |
| FDDI | Fibre distributed data interface |
| FDM | Frequency division multiplexing |
| FTP | Foil screened twisted pair |
| GHz | Gigahertz |
| HPPI | High performance parallel interface |

| | |
|---|---|
| HVAC | Heating, ventilation and air conditioning |
| ICEA | Insulated Cable Engineers Association |
| IEC | International Electro Technical Commission |
| IEE | Institute of Electrical Engineers (UK) |
| IEEE | Institute of Electrical and Electronic Engineers (USA) |
| ISDN | Integrated services digital network |
| ISO | International Standards Organisation |
| LAN | Local area network |
| LED | Light emitting diode |
| MAN | Metropolitan area network |
| Mb/s | Megabits per second |
| MHz | Megahertz |
| MUTOA | Multi-user telecommunications outlet assembly |
| nm | Nanometres |
| NEC | National Electrical Code (USA) |
| NEMA | National Electrical Manufacturers Association (USA) |
| NEXT | Near end crosstalk |
| NIC | Network interface card |
| NRZ | Non-return to zero |
| PAM | Pulse amplitude modulation |
| PAN | Personal area network |
| POF | Plastic optical fibre |
| PS-ACR | Power sum attenuation to crosstalk ratio |
| PS-ELFEXT | Power sum equal level far end crosstalk |
| PS-NEXT | Power sum near end crosstalk |
| SAN | Storage area network |
| SCSI | Small computer system interface |
| SFF | Small form factor (optical connectors) |
| SNR | Signal to noise ratio |
| SONET | Synchronous optical network |
| STP | Screened (or shielded) twisted pair |
| ScTP | Screened twisted pair |
| S-FTP | Screened, foil twisted pair |
| TDM | Time division multiplexing |
| TIA | Telecommunications Industry Association |
| TO | Telecommunications outlet |
| TOC | Terminated open circuit |

| TSB | Telecommunications Systems Bulletin (Technical Services Bulletin) |
| UTP | Unscreened (or unshielded) twisted pair |
| VCSEL | Vertical cavity surface emitting laser |
| WAN | Wide area network |
| WDM | Wavelength division multiplexing |
| xDSL | Digital subscriber link ('x' can denote different methods) |

# 1

# Introduction

This book is written as an aid to understanding the complexities of structured or premises cabling systems that are used to make today's office-based local area networks (LANs) actually work. So often we hear statements to the effect that cabling only accounts for 0.5% of the company's information technology budget yet accounts for 30% of the IT problems! Cabling may not be the most glamorous subject but the cables represent the foundation, indeed the *physical layer*, of the IT infrastructure. If the cabling isn't designed, procured, installed and tested correctly then the top six layers of the OSI networking model will simply be unable to function. It is probably fair to say that the cabling, despite its operational importance, capital outlay and long life expectancy, does indeed only attract 0.5% of management time. This book will hopefully help people to redress that balance.

As ever we are living through times of great change. The inexorable rise in the power of the microprocessor leads to faster and more powerful personal computers and the software available always seeks to stretch the processing and memory capabilities of the machine even further. To make full use of that processing power, computers need to communicate at faster and faster speeds. According to 'Amdahl's law' a megabit per second of input/output capability is needed for every MIPS (million instructions per second) of processing power. The new millennium will see the introduction of 64-bit microprocessors running at 1000 MHz clock speeds with hundreds of MIPS capability. 'Thanks, but I have enough bandwidth already', is guaranteed to be the most unlikely statement to be pronounced by anybody involved in information technology in the twenty-first century.

The 1970s saw data speeds topping out at 19.6 kb/s. LANs arrived with 10 Mb/s Ethernet and 4 Mb/s in the 1980s. The 1990s saw Fast Ethernet at 100 Mb/s and ATM at 155 Mb/s. The end of the 1990s saw the introduction of gigabit Ethernet and fibre channel; ten-gigabit Ethernet is planned by 2003. Thus communications speeds are increasing by a factor of ten every five or six years.

Structured cabling standards have struggled to keep up. Most of the familiar 'category 5' standards in use today have been fairly stable since 1995, but the advent of gigabit Ethernet over copper cable in the summer of 1999 demonstrated that the era of 'old' category 5 was over and enhanced category 5, or cat5e, standards started to make an entrance around the end of 1999. 2001 will see the arrival of category 6 and category 7 cabling standards and probably a new family of optical fibres to cope with the demands of ten-gigabit Ethernet over 300 m links. Chapter 15 however explores the wider concepts and applications of standards and demonstrates that it is wrong to get obsessed with 'categories' of cabling to the exclusion of all else. The wider framework of standards encompasses the requirements of EMC regulations, fire safety regulations, component specification, testing standards and the needs and expectations of LANs.

Apart from use as a formal textbook it is also presented as an aid for IT managers, consultants, cable installation engineers and system designers who need to understand the technology and physics behind the subject and the vast panoply of standards that accompany it. The book does not present itself as a design manual for structured cabling but rather explains the terminology and physics behind the standards, what the relevant standards are, how they fit together, and where to obtain them from. Anybody studying this book will be able to read the standards, understand manufacturers' data sheets and their conflicting claims and be suitably equipped to address those problems raised by the need to design, procure, install and correctly test a modern cabling system, using both copper and optical fibre cable technology.

It is presumed that the reader will be educated in mathematics and physics to a level somewhere equivalent in the UK to GCSE and in the US to 11th grade, but wherever the mathematical ability required extends beyond arithmetic, such as in the understanding of logarithms and decibels, then it is fully explained in chapter two, *Basic applied mathematics*. Similarly, fundamental physics issues are covered in chapters three and four, such as the laws of refractive index, basic electrical properties etc, permitting a full understanding of the later chapters devoted to copper and optical cable engineering. Hopefully the dedicated reader will find the book sufficiently self-contained to explain the subject in adequate, but not excessive, detail to meet their professional needs in this field.

# 2

# Basic applied mathematics

There are many excellent books on physics, maths and applied maths, suitable for all levels, and it is not the intention to reproduce large tracts of maths and physics here. There are certain fundamentals however that are necessary to adequately understand the process of selecting, designing, procuring, installing and testing a cable system. Some of these basic elements, such as decibels, occur over and over again in this particular subject of cable engineering, and it is topics such as these that will be explored in this chapter to give the reader a sufficient foundation of knowledge to make best use of the remaining chapters.

## 2.1 Working with indices

To handle very large or very small numbers, we use a convention of representing these numbers in the following notation, for example:

$$10^9 = 1\,000\,000\,000$$
$$10^{-3} = 0.001$$

The small number in the superscript, or the index, if it is positive, represents how many zeroes there are, or more precisely, how many factors of ten are involved. So $10^9$ means one thousand million, or a billion.

If the index is negative, then the number is less than one, and the index number reveals how many places after the decimal point there should be, or how many factors of divisions of ten there are. So $10^{-3}$ means one-thousandth. An expression of $6.3 \times 10^6$ means $6\,300\,000$.

If two numbers in this notation are multiplied together then simply add the index numbers together. For example:

$$10^4 \times 10^7 = 10^{11}$$
$$10^{-3} \times 10^7 = 10^4$$
$$(2 \times 10^4) \times (8 \times 10^6) = 1.6 \times 10^{11}$$

To divide two numbers, subtract one index from the other. For example:

$$10^9 / 10^6 = 10^3$$
$$10^{12} / 10^{-3} = 10^{15}$$
$$8 \times 10^8 / 2 \times 10^4 = 4 \times 10^4$$

Simple addition or subtraction of two numbers in this form can only take place if the indices are the same. For example:

$$(8.5 \times 10^6) - (3.1 \times 10^6) = 5.4 \times 10^6$$

# 2.2   Prefixes to denote size

There are accepted prefixes we can use to denote the size of a number more simply than always writing it out or pronouncing it in its entirety, e.g. *kilo*metres means one thousand metres; the *kilo* part representing one thousand. Table 2.1 gives the full list. For example:

1 pF is one picofarad, or $1 \times 10^{-12}$ farads.

Table 2.1 Prefix notation

| Decimal number | Index | Prefix | Symbol |
|---|---|---|---|
| 1 000 000 000 000 000 | $10^{15}$ | peta | P |
| 1 000 000 000 000 | $10^{12}$ | tera | T |
| 1 000 000 000 | $10^{9}$ | giga | G |
| 1 000 000 | $10^{6}$ | mega | M |
| 1 000 | $10^{3}$ | kilo | k |
| 100 | $10^{2}$ | hecto | h |
| 10 | $10^{1}$ | deca | da |
| 0.1 | $10^{-1}$ | deci | d |
| 0.01 | $10^{-2}$ | centi | c |
| 0.001 | $10^{-3}$ | milli | m |
| 0.000 001 | $10^{-6}$ | micro | $\mu$ |
| 0.000 000 001 | $10^{-9}$ | nano | n |
| 0.000 000 000 001 | $10^{-12}$ | pico | p |
| 0.000 000 000 000 001 | $10^{-15}$ | femto | f |
| 0.000 000 000 000 000 001 | $10^{-18}$ | atto | a |

# 2.3   Logarithms

Logarithms of numbers also make the large and the small easier to manipulate. A logarithm of a number is that number that you have to raise another number to the power of to get back to the first number. For example:

The logarithm of 100 is 2, because you have to raise 10 to the power of 2 ($10^{2}$) to get 100.

This is working to base ten, but logarithms can be expressed in any base. To be precise we should write $\log_{10}$, but it is always assumed that if no base number is specified then we are talking about calculations made to the base ten. These are sometimes referred to as common logarithms. Sometimes base 2 is used in communications theory because digital transmission only has two states, that is 'ones' and 'zeroes'. The $\log_{2}$ of 16 is 4, because we have to raise 2 to the power of 4 ($2^{4}$) to get 16.

Antilogarithms are simply working the other way round. For example:

antilog$_{10}$2  = 100
antilog$_{10}$3  = 1000
antilog$_{10}$−3 = 0.001

# 2.4   Decibels (dB)

Decibels are ten times the logarithm of a ratio. They are used in all branches of engineering, and can be used to represent differences in electrical power, light or even sound. Using decibels makes calculations much easier to comprehend and even do in your head. For example, if the attenuation of one piece of cable is 4 dB, and you add onto it another length of cable with 3 dB of attenuation then the resulting attenuation of the whole channel is 7 dB. Decibels are a convenient shorthand which show how energy is absorbed or produced regardless of what levels of energies are actually involved:

gain (attenuation if it is negative) = $10\log_{10}(P_1/P_2)$,                    [2.1]

where $P_1$ is one power measurement and $P_2$ is another power measurement that we wish to compare with the first.

For the remainder of this book we will adopt the convention that all logarithms are to the base 10 unless otherwise denoted. For example:

if power level 1 (output) equals 1 mW and power 2 (input) equals 0.002 mW then

gain = 10 log 1/0.002

    = 10 log 500

    = 10 × 2.7 = 27 dB

for attenuation we have

10 log 0.002        ($10\log P_2/P_1$)

= −27 dB

Note the absolute value does not change, only the sign. Many writers leave out the negative sign altogether if it is clear they are talking about attenuation, so as to avoid the uncertainty of a double negative.

Sometimes it is presumed that the comparison is being made to 1 mW of power, so $P_1$ or $P_2$ in the equation will always be one. To denote this the resulting answer has the units dBm.

Measuring the power in a device or a cable is not nearly as easy as measuring the voltage across it relative to ground or any other potential. We can take account of this by knowing that:

power = volts × amps, or $P = V \times I$  [2.2]

But from Ohms law we also know that $V = IR$ and hence $I = V/R$ and therefore power also equals $V \times V/R$ or $V^2/R$.

$R$ is the resistance of the cable and is presumably the same from one reading to another, so we can cancel it out in the following equation:

$$
\begin{aligned}
\text{gain} &= 10\log\left(\frac{V_1^2/R}{V_2^2/R}\right) \\
&= 10\log(V_1/V_2)^2 \\
&= 20\log(V_1/V_2)
\end{aligned}
$$  [2.3]

So we can obtain decibel readings by simple voltage measurements and incorporating a factor of 2 in the standard decibel equation.

It should be remembered that power only equals voltage multiplied by the current for the special case of direct current. For alternating current the correct formula is:

power (W) = voltage × current × the cosine of the phase
difference between the two.

The cosine of the phase difference is known as the power factor. If the current and voltage are in phase then the angle is zero and the cosine of zero is one, so in that special case power does indeed equal voltage times current. Any reactive load, i.e. capacitance or inductance, will cause the current to lead or lag the voltage in phase.

## 2.5   Sine waves and phase

A sine wave or sinusoidal wave is the most natural representation of how many things in nature change state. A sine wave shows how the amplitude of a variable changes with time. The variable could be audible sound for example. A single pure note is a sine wave, although it would sound a very plain and flat note indeed with none of the harmonics we normally hear in nature. A straightforward oscillating or alternating current or voltage within a wire can also be represented by a sine wave. The number of times the sine wave goes through a complete cycle in the space of 1 second is called the frequency. Indeed the unit used to be cycles per second, but now the unit of measurement is hertz (Hz). A frequency of 1000 Hz, or 1 kHz, means that the sine wave goes through 1000 complete cycles in 1 s. If we are considering audible sound waves then the human ear has a frequency range of about 20 Hz–20 kHz. The electrical mains frequency in Europe is 50 Hz and 60 Hz in America. Figure 2.1 shows a sine wave.

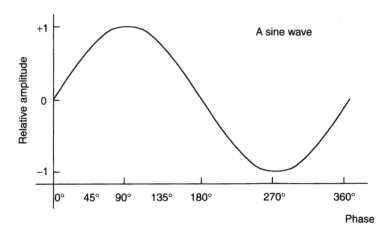

**Fig. 2.1** A sine wave.

The sine of any angle can vary from −1 to +1. For example the sine of 0° is 0 and the sine of 90° is 1. The sine of 270° is −1 and when we get to 360° we are back to zero again. A cosine is 90° out of phase with a sine wave as we can see in Fig. 2.2.

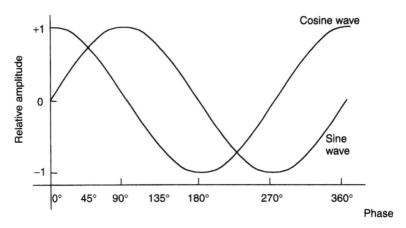

**Fig. 2.2** A sine and cosine wave.

The cosine of 0° is 1 and the cosine of 90° is 0. So we say that a cosine is 90° out of phase with a sine wave. Any number of sine waves can exist at any one time and have any manner of angular phase differences to each other. Whenever a phase angle is mentioned it is always relative to something else. Digital 'ones' and 'zeroes' can be encoded as two signals of identical amplitude and frequency but with different phases to each other or some other reference marker. This would be called phase modulation.

When we apply an alternating voltage across a resistor, a current flows through the resistor. If we looked at the voltage and current waveforms on an oscilloscope we would see two sine waves that superimpose each other, when differences of amplitude are taken into account. The two signals are *in-phase* with each other. If we add a capacitor in series with the resistor we would see the current and voltage signals diverge so they were out of phase with each other. When an electrical current flows in a circuit we are observing the effect of the flow of the fundamental particles called electrons flowing through the wire from a negative to a positive terminal. We can imagine a capacitor is like a big bucket for electrons. When the voltage is applied to the circuit, the electrons flood into the bucket. But as the rising voltage reaches its peak, the bucket is nearly full, and the flow of electrons, or current tails off. The flow of current therefore seems to lead the voltage and is out of phase with the voltage. An inductive load works the other way round. The rising voltage is needed to draft the electrons into the inductor where they fight against the magnetic field they have created. The current therefore lags the voltage. A circuit with a capacitive and/or inductive load is called reactive. The actual phase of the current relative to the voltage will depend on the values of resistance, capacitance and inductance in the circuit and may be represented as a *complex* number.

## 2.6   Complex numbers

Complex numbers take the form $a + jb$, where $a$ is the real number and $jb$ is the imaginary number. $j$ is supposed to be the square root of minus one, that's why it's called imaginary (just try getting an answer for the square root of minus one on your calculator and you'll

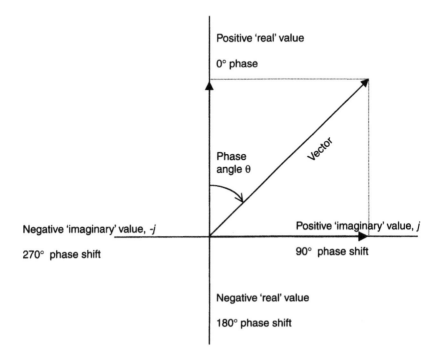

**Fig. 2.3** Complex vector.

see why). This strange format is useful to express a value which has a phase and amplitude component. The real component can represent the in-phase component, such as the current flowing through a resistor, and the imaginary component represents the current flowing out-of-phase due to the reactive load. As we can see from Fig. 2.3 the resulting value is a vector which has an amplitude and a phase angle caused by the complex combination of two or more out-of-phase components. Note that most books on mathematics use lower-case $i$ to represent the imaginary component. However in electrical engineering, $i$ more often means a current, so the letter $j$ is used instead.

# 3

# Basic physics — electrical

## 3.1  SI system and fundamental units

As long as people have been measuring things they have needed units of measurement to make any kind of meaningful recording of the event. Noah would not have understood God's instructions to make the Ark unless both understood the concept of the cubit. As empires have risen and fallen they have introduced their own units of measurement to make sense of their own mathematics, civil engineering and commerce. Today the world has evolved towards the metric system and away from a hotchpotch of measurement units vaguely grouped together under the heading of 'imperial'. Only the United States, the United Kingdom and Ireland use imperial measurements, although frequently mixed with metric units. The rest of the world is solidly metric! The number of imperial units in common circulation has declined and the era of British schoolchildren, pre-1970, having to struggle with rods, chains, pecks and bushels is fortunately over. Imperial units still hold some stings however; the Irish acre is not the same as the English acre, the US gallon and British gallon have little in common and the meaning of the nautical mile is open to interpretation!

Getting units wrong can be expensive. The Mars Climate Orbiter spacecraft crashed into the surface of Mars in 1999 because one programmer in NASA had been calculating in metric units whilst another had been using imperial units.

Whilst metric units are universally accepted for engineering and science they are more accurately described under the SI system, or *Système Internationale d'Unités*, which originated in 1948 and is now enshrined as an ISO standard. SI contains seven base units and some derived and supplementary units. The seven base units are:

1   The metre, m, the unit of length.
2   The kilogram, kg, the unit of mass.
3   The second, s, the unit of time.
4   The ampere, A, the unit of electric current.
5   The kelvin, K, the unit of temperature.
6   The candela, cd, the unit of luminous intensity.
7   The mole, mol, the standard amount of a substance.

All other units can be derived from these base units, e.g. the unit of force is the newton, N, but it can also be expressed as *mass* times *length* over *time*-squared. The SI unit of pressure is the pascal, Pa, this is force per unit area and so can be expressed as *mass* over *length* times one over *time*-squared. The SI unit of frequency is the hertz, Hz, which is the reciprocal of time.

Supplementary units include the radian and the steradian. Before the SI system some countries used the cgs system, meaning the basic units were centimetres, grams and seconds rather than the SI units of metres, kilograms and seconds. Other units still in popular use, but not recognised SI units, are the micron ($10^{-6}$m), the metric tonne (1000 kg) and minutes, hours, days and years.

In electrical engineering the base unit is the ampere. One amp is defined as that constant current which, if maintained in each of two infinitely long straight parallel wires of negligible cross-section placed 1 m apart, in a vacuum, will produce between the wires a force of $2 \times 10^{-7}$ N/m length.

From this we have the potential difference, whereby 1 V is the difference in electric potential between two points of a wire carrying a constant current of 1 A when the power dissipation between these two points is 1 W.

One ohm of resistance is defined as the electrical resistance between two points of a conductor when a constant potential differ-

| Table 3.1 SI electrostatic and electromagnetic units | | | |
|---|---|---|---|
| Quantity | Symbol | SI unit | Abbreviation |
| Mass | m | kilogram | kg |
| Length | l | metre | m |
| Time | t | second | s |
| Current | I | ampere | A |
| Charge | Q | coulomb | C |
| Potential difference | V | volt | V |
| Power | P | watt | W |
| Resistance | R | ohm | $\Omega$ |
| Conductance | G | siemens | S |
| Inductance | L | henry | H |
| Capacitance | C | farad | F |
| Magnetic flux | $\Phi$ | weber | Wb |
| Magnetic induction | B | tesla | T |
| Electric field strength | E | volt metre$^{-1}$ | Vm$^{-1}$ |

ence of 1 V applied between these two points produces in the conductor a current of 1 A.

The full list of SI units relevant to electrical engineering is in Table 3.1.

# 3.2    Atoms, elements and compounds

The atom is the basic building block of matter in which the matter still retains the unique identity of an element. Copper is an element and it is made up of copper atoms. Atoms themselves are made up of building blocks such as electrons and protons, but if the copper atom is reduced to its constituents then it is no longer a copper atom in the same way as if a wool coat was unravelled into a ball of wool, cotton and buttons, the identity of the original coat would be lost, and those constituents could go to make something else.

Elements are made up of atoms of the same type. Iron, copper, aluminium are all elements; steel is not, it is a mixture of iron and carbon and a few other things.

Molecules are groups of atoms. Oxygen atoms, for example, go around in pairs, so the atmosphere is full of oxygen molecules, which have the symbol $O_2$. Compounds are two or more elements that are chemically combined, such as water, $H_2O$, which is two atoms of hydrogen and one of oxygen. Salt is sodium chloride, NaCl, one atom of sodium (Na) and one of chlorine, (Cl).

Allotropes are ways in which the same element can exist in different forms. For example, both diamond and graphite are allotropes of carbon, as is fullerene.

There are 92 elements found on earth and a further 14 have been produced by scientists.

Atoms are made up of three basic subatomic particles, called protons, neutrons and electrons. In the last 30 years 'atom-smashers', or high velocity particle colliders, have broken atoms into a large and bizarre list of sub-subatomic particles. However we merely need to grasp the fundamentals of the proton, neutron and electron. An atom is made up of a nucleus containing protons and neutrons. Around this in orbit are electrons that group together in various shells. Figure 3.1 gives a representation of an atom.

A proton has one positive charge and an electron has one negative charge. The proton is many times more massive than an electron however. The neutron has no charge but the same mass as the proton. The atom has a neutral overall electric charge because there are as many electrons as there are protons. If an atom loses or gains electrons then it would become charged and would be known as an ion. The process of losing electrons is called ionisation.

The mass of an atom is made up of the combination of neutrons and protons, the electrons add very little to the mass. The number of protons and neutrons together is the mass number. The number of protons is called the atomic number. Most of the volume of an atom is just empty space. When some stars collapse at the end of their life to form neutron stars the atoms have their electrons crushed down towards the nucleus by immense gravity and pressure. This is why neutron star material has unimaginable weight such as around a million tonnes per spoonful!

The electrons orbit in shells. One or two in the first shell, up to 8 in the second shell, up to 18 in the third shell (an inner group of 10 and an outer group of 8) and up to 32 in the outer shell (including an outer group of 8). Electrons fill up the lower shells first and each shell

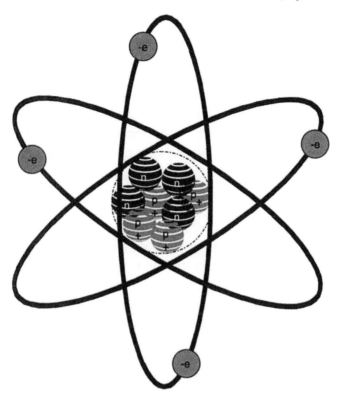

**Fig. 3.1** The atom.

is associated with a particular energy level. Phosphorescent materials allow photons of light to boost some of the electrons to higher levels. When the electron falls back a level it releases a photon to account for the energy, and that is what we would see as light.

Hydrogen is the simplest element. It has one proton and one electron and zero, one or two neutrons depending upon which isotope of hydrogen it is. Next is helium with two of everything. The more subatomic materials in an atom the heavier and more dense it will be. Uranium has a mass number of 235. The number of electrons in the outer orbits determines the chemical properties of an element.

All the members of the group of elements known as the halogens (fluorine, chlorine, bromine, iodine and astatine) have seven electrons in their outer shell so they all react very similarly. A group of gasses known as the noble gasses have all of their outer electron shells filled up, so they are extremely unreactive. This group includes helium, neon, and argon.

The elements are usually listed in something called the periodic table. Elements, which have the same number of electrons in the outermost shell, fall into vertical columns, of which there are eight. The horizontal listings are called periods, of which there are seven plus two 'rare earth' periods called the lanthanum series and the actinium series. The latter two periods contain the mostly artificially produced and very unstable elements such as curium, plutonium and californium etc.

In between the very reactive metals such as lithium and sodium which lie on the left side of the periodic table, and the non-metals such as chlorine and fluorine, on the right of the table, lie the transition metals, and here we see the more familiar metals such as iron (Fe), copper (Cu) and gold (Au). At the border of the metals and non-metals we find the metalloids which have some metallic and some non-metallic properties. Here, for example, we find silicon and germanium.

To be an atom of the same element it must have the same number of protons as every other atom of the same element. So, for example, every single atom of magnesium must contain 12 protons, or else it wouldn't be a magnesium atom. The number of neutrons however may differ. Forms of an element which differ in the number of neutrons in the atom are called isotopes. Isotopes are expressed in the form $^A_ZX$, where $X$ is the chemical symbol, $A$ is the mass number (neutrons plus protons) and $Z$ is the atomic number, i.e. the number of protons. For example, two isotopes of chlorine are $^{35}_{17}Cl$ and $^{37}_{17}Cl$.

Carbon, for example, has three isotopes, known as carbon-12, carbon-13 and carbon-14. The latter is radioactive, with a half-life of 5700 years, and is used to 'carbon date' material that was once alive, typically wood.

# 3.3   Conductors, semi-conductors and insulators

An electrical current flows in a conductor when a potential difference, i.e. a voltage, is applied across it. The flow of current is the flow of electrons within the conductor moving under the influence of the applied electric field. Metals are conductors and copper is a very good conductor. The atoms within a metal are held together by what is known as the metallic bond. The outer electrons of the atoms are able to break free and roam about between the metal ions. The motion of the electrons is random except when under the influence of an electric field.

Not all metals can conduct electricity to the same extent. Copper is much better than aluminium for example, with copper having only two-thirds the resistivity of aluminium.

An insulator is a material within which the electrons are securely fixed in chemical bonds and are not free to move about under the influence of an electric field. Eventually though, when the voltage was high enough, a current would flow through the material and that is called the breakdown voltage. Insulators are also sometimes known as dielectrics. An insulator, or dielectric, has the capacity to store a charge to some extent as the atoms are pulled apart slightly by the electric field. It is a store of energy in a similar way that a stretched elastic band is a store of potential energy. This capacity is known as the material's permittivity. A vacuum could not hold any charge because it contains no matter. A material is thus measured by its relative permittivity, i.e. how much better is it at storing a charge than a vacuum.

A capacitor is two conductors separated by a dielectric. If the capacitance of the two plates is measured in dry air, then the relative ability to store a charge is measured when different materials are inserted between the two plates, and we can then determine the relative permittivity of that material. The relative permittivity of air is 1.000536, because air is not that far removed from a vacuum, but the relative permittivity of mica is 6.7, so a piece of mica between two plates of metal would make a far better capacitor than two parallel metal plates on their own. Capacitance is also proportional to the surface area of the two conductors and inversely proportional to their distance apart.

Capacitance is an unwelcome phenomenon in cables. The capacitance takes time and energy to charge and discharge as high-speed electronic signals pass down the wire. It also provides a mechanism whereby noise can be coupled into the cable.

Conductors have the property of inductance, which has the symbol, L, and the unit of the henry, H. Inductance is the ability of a conductor to store a magnetic field. Like capacitance, inductance consumes time and energy from the signal by having to charge and discharge the magnetic field.

Inductance is a property of the conductor, whereas capacitance has to be between two conductors or more. We can thus have capacitance between conductors in the same cable, between the conductors and a screen or shield, and capacitance between the conductors and the ground plane. The properties of a cable can thus be changed by the cable's proximity to an earthed/grounded surface.

Finally we have semiconductors, such as silicon and germanium. Semiconductors have properties between metals and non-metals and have proved invaluable in the development of semiconductor microcircuits based primarily on silicon.

# 3.4   Electricity and circuits

We have seen that electrical current is the flow of electrons through a conductor under the influence of an electric field. If the voltage stays at the same polarity then the flow of current will always be in one direction. This is called direct current, or dc. If the polarity of the voltage changes then so will the current. This is called alternating current, or ac. The domestic voltage system we use in our houses and in the national grid is ac, primarily because it makes it easy to change from one voltage to another by the use of transformers.

The main parameters used to quantify an electrical circuit are:

- voltage
- current
- resistance
- capacitance
- inductance
- impedance

We also have conductance, G, which has the unit of siemens, conductance is the reciprocal of impedance or resistance and may be considered as the leakage between conductors due to imperfect insulation.

For a dc circuit we can relate the resistance, voltage and current by Ohm's law, given in equation 3.1.

$$V = I \times R \tag{3.1}$$

$V$ = voltage, in volts
$I$ = current, in amperes
$R$ = resistance, in ohms.

For an ac circuit the resistance to the current is the combined effect of the dc resistance, and the reactance ($X$) of the capacitance and the inductance. The effect is called the impedance, and it has the symbol, $Z$. The units are still in ohms. Equation 3.2 gives the relationship of $R$ and $X$ to create the ac impedance:

$$Z = R + jX \tag{3.2}$$

$Z$ = impedance, in ohms
$R$ = resistance, in ohms
$X$ = reactance of the circuit, in ohms.

Reactance is the complex (in the mathematical sense, hence the operator '$j$' in front of it to denote a complex number. Pure mathematicians would use the symbol '$i$' as the complex number operator, but to engineers this more often means current and so engineers usually use the symbol '$j$') value of impedance inherent in a capacitor and inductor. It is frequency dependent and the complex operator denotes that it has a phase value. Equation 3.3 gives the reactance of a capacitor:

$$X_c = 1/j\omega C \tag{3.3}$$

$X_c$ = reactance, in ohms
$C$ = capacitance, in farads
$\omega$ = angular frequency, i.e. $2\pi f$, where $f$ = the frequency in hertz.

Equation 3.4 gives the reactance of an inductor:

$$X_L = j\omega L \tag{3.4}$$

$X_L$ = reactance, in ohms
$L$ = inductance, in henries
$\omega$ = angular frequency, i.e. $2\pi f$, where $f$ = the frequency in hertz.

Resistances and impedances, when in series, can simply be added together to give the circuit total, i.e. $R_t = R_1 + R_2 + R_3 + \ldots$ When in parallel, the total value becomes $1/R_t = 1/R_1 + 1/R_2 + 1/R_3 + \ldots$ Figure 3.2 shows this.

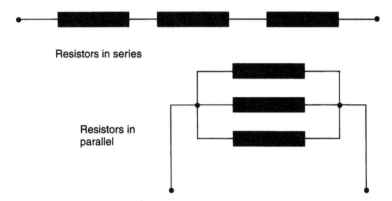

Resistors in series

Resistors in parallel

**Fig. 3.2** Resistors in series and parallel.

Of more use in communications is the characteristic impedance of a circuit, rather than the absolute impedance. The characteristic impedance, $Z_0$, is the impedance of an infinite length of line. Even though real-life cables are never infinitely long the characteristic impedance is important for matching components in a circuit together. Energy can only be transferred from the source to the cable and onto the load efficiently, if the characteristic impedances of each component are the same. Otherwise energy will be reflected back at every point where there is a discontinuity of characteristic impedances. Equation 3.5 gives the equation for characteristic impedance in ohms:

$$Z_0 = \sqrt{\frac{R + j\omega L}{G + j\omega C}} \qquad [3.5]$$

We may assume that the conductance, $G$, goes to zero for a cable with a good insulator. Also at high frequencies $R$ will become very small compared to the $\omega L$ term and so the equation simplifies to:

$$Z_0 = \sqrt{\frac{j\omega L}{j\omega C}}$$ [3.6]

And then to:

$$Z_0 = \sqrt{\frac{L}{C}}$$ [3.7]

as the $j\omega$ terms cancel out.

The amount of energy reflected back can be derived from:

$$R_s = \frac{Z_s - Z_0}{Z_s + Z_0}$$ [3.8]

where $R_s$ = the reflection coefficient (also given the symbol $\rho$ in some textbooks)

$Z_s$ = the impedance of the source
$Z_0$ = the characteristic impedance of the cable.

From equation 3.8 we can see that the reflection coefficient will be zero when $Z_S = Z_0$

return loss = $20 \log_{10} R_s$ [3.9]

A twisted pair telephone cable can be represented as two resistors in series with a capacitor in parallel with one of them. The capacitor appears as an open circuit at dc and tends towards a short circuit as the frequency gets higher thus shunting out the effect of the second resistor; see Fig. 3.3.

We can deduce from the diagram that the impedance of the line is around 600 ohms at low frequencies and tends towards 100 $\Omega$ at

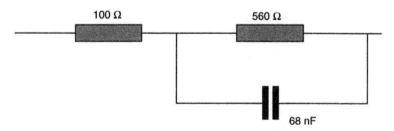

**Fig. 3.3** Equivalent impedance model for a twisted pair, 0.5 mm copper.

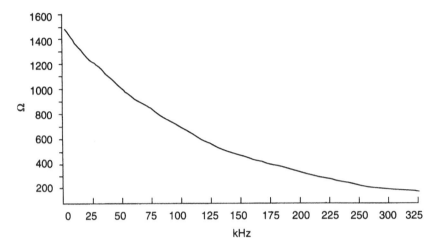

**Fig. 3.4** Impedance against frequency for a 0.5 mm copper pair.

higher frequencies. A standard test on a telephone line is to look for 600 Ω impedance at 1200 Hz. Figure 3.4 shows the impedance of a 0.5 mm copper pair against frequency. At audio frequencies, i.e. 30 Hz–3.3 kHz the impedance is of the order of 1400–600 Ω, but at higher frequencies, and remember that Local Area Networks will generally be operating in the 10–100 MHz band, the impedance tends towards 100 Ω.

Electrical power is measured in watts. Energy is power times time. Hence we pay our electricity bills in units of kilowatt-hours, being the amount of power we are consuming times the period that consumption goes on for. If we run a 2 kW electric fire for 3 hours then we will have consumed 6 kW-hours of energy. Although kW-hours (or simply 'units' as they may be referred to on your electricity bill) is a convenient unit of measurement for a power company, the correct SI unit of energy is the joule, J, or kilojoule, kJ.

We can derive the power consumption in watts of a dc electrical circuit from the following equations:

$$\text{power} = V \times I \qquad\qquad [3.10]$$
$$\text{power} = I^2 \times R \qquad\qquad [3.11]$$
$$\text{power} = V^2/R \qquad\qquad [3.12]$$

For an alternating current circuit we can replace the resistance by a value for the impedance, but equation 3.10 then requires modification. In a purely resistive circuit the voltage has the same phase as the current. If we envisage a sinusoidal voltage then the resulting current is exactly the same shape and the peaks and troughs of the two waveforms are exactly coincident, i.e. they are in phase. If the load is reactive, i.e. it has capacitance and/or inductance then the current will not be in phase with the voltage. If, for example, we apply a voltage across a capacitor, the current will rapidly flow into the capacitor as electrons seek to 'fill-up' the bucket they see before them. The voltage will rise slowly as the current flows but as the bucket of electrons fills up the current will slow down and the voltage will rise to its maximum potential. If the two waveforms, current and voltage, were observed on an oscilloscope we would see the current apparently leading the voltage wave by 90°. The opposite will happen with an inductor. If we tried to arrive at the power generated by multiplying the current by the voltage we would not get a correct answer as when the voltage was at a maximum, the current flow would be zero. Power dissipation in an ac circuit is given in watts by equation 3.13.

$$\text{power} = V \times I \times \cos\theta \qquad\qquad [3.13]$$

$\cos\theta$ = the cosine of the phase angle between the voltage and current.

Cos $\theta$ is also known as the power factor. The cosine of 90° is 0, so that in the worst case, with 90° phase lag, no power would be dissipated. To avoid confusion or ambiguity many machines and generators quote their output or consumption in terms of kVA, or kilovolt-amperes, demonstrating that absolute power dissipated depends upon the reactance of the load applied.

# 4

# Basic physics — optical

## 4.1  The electromagnetic spectrum

The electromagnetic spectrum demonstrates how different forms of electromagnetic radiation have different wavelengths and it is the wavelength that distinguishes one form of radiation from another. At long wavelengths we have radio waves and at the other extreme we have gamma rays of very short wavelength. Figure 4.1 shows the electromagnetic spectrum.

When travelling in free space electromagnetic energy also has the nature of particles, which are called photons.

The wavelength is related to the frequency by the following formula,

$$f = c/\lambda \qquad\qquad [4.1]$$

$f$ = frequency, in hertz
$\lambda$ = wavelength, in metres
$c$ = velocity of light, in m/s.

In the radio part of the spectrum we have high frequency (HF) and very high frequency (VHF) bands which move up in frequency, and down in wavelength as we approach the microwave range:

- low frequency (LF)            300 kHz–3 MHz    (1000–100 m)
- high frequency (HF)           3–30 MHz         (100–10 m)
- very high frequency (VHF)     30–300 MHz       (10–1 m)
- ultra high frequency (UHF)    300–3000 MHz     (1–0.1 m)

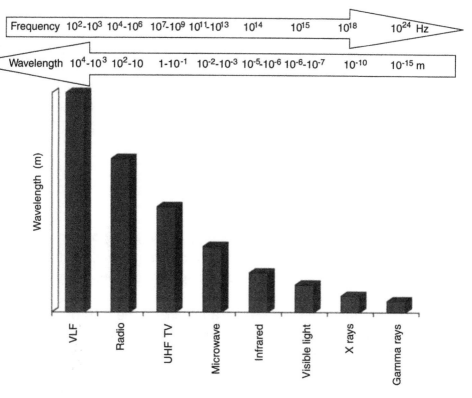

Fig. 4.1 The electromagnetic spectrum.

Radar works in the 10 cm–10 mm band. We reach the infrared band at wavelengths around $10^{-4}$–$10^{-6}$ m, and at this point it is more convenient to move into the units of nanometres to describe them. A nanometre (nm) is a thousand-millionth of a metre or $10^{-9}$. 1000 nm is $10^{-6}$ m. An older unit sometimes used is the Ångstrom which is $10^{-10}$ m.

Visible light is in the band 400–700 nm, violet is around 400 nm and red around 600–700 nm. Yellow is from 570–590 nm. The human eye has the highest sensitivity at this wavelength as it matches the peak output power of the sun in the visible light range.

Past visible light we move into ultra-violet and then onto X-rays at $10^{-12}$ m and then onto gamma rays at $10^{-20}$ m. Radiation from ultra-violet (UV) and beyond is sometimes referred to as ionising radiation as it has the energy to strip electrons away from atoms. This can make it very injurious to animal and plant life. The atmosphere of our planet protects us from the majority of ionising radiation coming from the sun.

The infrared light sources used in fibre optics are outside the range of the human eye and that is why many pieces of optical equipment carry warnings about invisible laser radiation. Optical fibre sources work in the band 850–1550 nm.

# 4.2   Reflection and refraction

Light travels through a vacuum at a speed of 299 792 km/s, usually abbreviated to 300 000 km/s or more correctly $3.0 \times 10^8$ m/s. When light travels through another transparent or translucent material it is significantly slowed down.

The refractive index, $n$, of a material is the ratio of the velocity of light in free space to that in the material in question. The refractive index of glass is around 1.5. This means that light travels in glass at around 300 000/1.5 = 200 000 km/s. It is a little publicised fact that the speed of a signal in a copper cable is actually faster than the speed of a signal in an optical fibre, about 5% faster. Light goes a little faster in ice ($n = 1.31$), but slows down to a crawl through a diamond ($n = 2.4$).

The change in velocity as a light beam passes through a medium of one refractive index to another is what causes the phenomenon of refraction, or the apparent changing in angle of the incident light beam as it passes through the boundary of the two materials.

If the surface between the two media is purely reflective, and flat, then the incident light will reflect off at the same angle to the flat plane as the incident light, i.e. the angle of incidence equals the angle of reflection.

If the boundary is between two transmissive materials of different refractive index then light approaching below a certain angle of incidence will be reflected back off the boundary. At a certain angle the light will be refracted along the boundary layer. Above the angle the

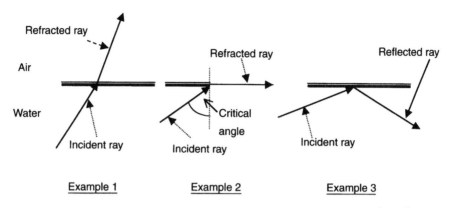

Fig. 4.2 The critical angle.

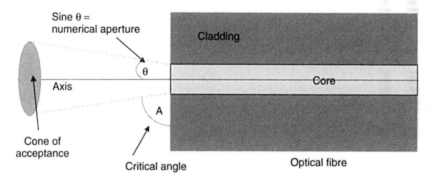

Fig. 4.3 Numerical aperture.

light will be refracted through the boundary. This angle is called the critical angle, see Fig. 4.2.

The angles of incidence (i) and refraction (r) are related by Snell's law, which states that the ratio of the sine of the incident angle to the sine of the refracted angle is equal to the ratio of the refractive indices of the two media. If the first media is air then the refractive index is one so that the refractive index of the second material becomes simply sin i/ sin r.

Optical fibre works by having a core of transmissive material of one refractive index surrounded by a cladding of another material with a lower refractive index. Light rays travelling through the core and hitting the core/cladding boundary below the critical angle will be reflected back into the core and constrained to stay within the core. This is called total internal reflection.

When light is launched into the core from outside, any part of it approaching the core/cladding above the critical angle will be lost through the cladding due to refraction. We can thus imagine a cone of acceptance being formed axial to the core. Light approaching the core within the cone of acceptance will propagate through the core. Light from outside the cone will be lost. The sine of the angle the cone makes with the axis of the core is called the numerical aperture, NA. The larger this figure is then the easier it is to launch light into the core. Optical fibres typically have numerical apertures of around 0.2, see Fig. 4.3.

# 5

# Communications theory

## 5.1 Analogue and digital channels

Communications theory considers the manner in which useful information is transmitted through a communications channel whilst subject to noise.

The *channel* is the communications medium in between the starting point, the transmitter, and the finishing point, the receiver. The channel, shown in Fig. 5.1, may be free space, as in radio transmission or a more tangible medium such as copper or optical cable. Exactly what the channel is made of is not particularly relevant to communications theory; the fundamental mathematics and problems remain the same. Every communications channel is subject to unwanted noise.

The *information* is the useful work we get out of the channel. It can be digital or analogue. *Digital* means that the information has been encoded into a binary series of 'ones' and 'zeroes', although some form of modulation may be added to the digital information to expedite its transit through the communications channel. *Analogue* means that the information is an *electrical or optical analogue*, i.e. a representation, of an original signal, that can have any one of an infinite number of values at any point in time.

Consider a Rolex wristwatch with a conventional sweep second hand. The second hand rotates continuously around the dial in a smooth motion. If you looked at the second hand at any

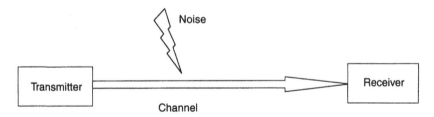

**Fig. 5.1** Communication channel.

random point in time it could be at any one of an infinite number of positions. Its value is therefore an analogue of time. If we look at a Seiko watch on the other hand (not literally, it could still be on your left wrist!) its second hand moves in a stepped manner from one second marker to another. If we ignore the small length of time to make the physical move from one marker to another, we know that if we look at the face of the wristwatch at any random moment in time the value will be in only one of sixty possible places. This is a digital representation of time. The hour and minute hands of both watches are still giving an analogue reading. A true digital watch or clock does of course show a numerical readout of time.

There can be no uncertainty of the time with a digital clock, you literally read the time to the nearest second (assuming it has an hour, minute and second display). But of course that is the point of uncertainty. If the clock says twenty-nine seconds past the minute, we know that the real time is somewhere between twenty-nine and thirty seconds past the minute. We don't know what it is until the readout changes from 29 to 30, but then the problem starts again! In communications theory this uncertainty is called *quantization noise*, which shows that the digital representation is not an absolutely accurate representation of our analogue reality. But for most practical purposes, it is close enough. So in all analogue to digital conversions we have to decide how much accuracy we can tolerate. The more accuracy we aim for then the more processing power, memory, but ultimately bandwidth, we require, and hence more cost.

If we encode, or represent a song with four-bit binary words, the resulting digital-to-analogue decoding would sound unacceptably

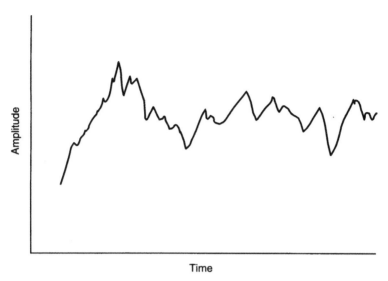

**Fig. 5.2** An analogue waveform.

poor to our ears. The quantization noise added to the original would be unacceptable. A CD player usually uses 16-bit words to encode high-fidelity music. This is a level of accuracy far beyond what the human ear could discern in the way of added quantization noise. Figure 5.2 shows an analogue waveform and Fig. 5.3 shows its 'sampled' version.

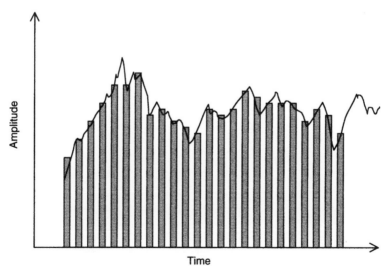

**Fig. 5.3** Waveform 'sampled' to digitise it.

## 5.2   Binary, octal and hexadecimal

### 5.2.1   Binary

Normally we count in base ten, i.e. 0,1,2,3,4,5,6,7,8,9 then we have to start again because we've got to ten, so we move onto 10,11,12,13 . . . and so on. There is nothing natural in mathematics about counting in base ten, it probably originated because we have ten fingers (if we include our thumbs) on both hands. The Romans also counted in base ten but got into all sorts of problems when they tried to write down long calculations because they never figured out the concept of 'zero'. Just try the following calculation:

(XXIV times IV) only using Roman numerals. Now do the equivalent sum in your head, 24 × 4.

Computers don't find counting in base ten particularly useful, or easy. A computer is essentially a set of switches that exist in one of only two states, on or off. They are digital machines that live in a binary world of great simplicity, either it is on, or off. Our base ten numbers therefore have to be converted to binary.

Any base ten number can be represented by a binary number of ones and zeroes. These are often grouped together in bundles of eight digits to form a byte, or digital word. An eight bit word can represent up to 256 states, with the most significant digit representing 128, then 64, then 32 and so on.

| Base ten numbers | | | Binary equivalents | | | | | | 4 bit binary word |
|---|---|---|---|---|---|---|---|---|---|
| 100 | 10 | 1 | 32 | 16 | 8 | 4 | 2 | 1 | |
| | | 0 | | | | | | 0 | 0000 |
| | | 1 | | | | | | 1 | 0001 |
| | | 2 | | | | | 1 | 0 | 0010 |
| | | 3 | | | | | 1 | 1 | 0011 |
| | | 4 | | | | 1 | 0 | 0 | 0100 |
| | | 5 | | | | 1 | 0 | 1 | 0101 |
| | | 6 | | | | 1 | 1 | 0 | 0110 |
| | | 7 | | | | 1 | 1 | 1 | 0111 |
| | | 8 | | | 1 | 0 | 0 | 0 | 1000 |
| | | 9 | | | 1 | 0 | 0 | 1 | 1001 |
| | 1 | 0 | | | 1 | 0 | 1 | 0 | 1010 |
| | 1 | 1 | | | 1 | 0 | 1 | 1 | 1011 |

## 5.2.2   Octal

The next stage may be to count in base 8, if groupings of eight are a useful way to represent data.

| Base 10 | | | Base 8 | | | Octal word |
|---|---|---|---|---|---|---|
| 100 | 10 | 1 | 64 | 8 | 1 | |
| | | 0 | 0 | 0 | 0 | 000 |
| | | 1 | 0 | 0 | 1 | 001 |
| | | 2 | 0 | 0 | 2 | 002 |
| | ................ | | | | | |
| | | 8 | 0 | 1 | 0 | 010 |
| | 1 | 0 | 0 | 1 | 2 | 012 |
| 1 | 3 | 2 | 2 | 0 | 4 | 204 |

### 5.2.3   Hexadecimal

A 16-bit word can be represented using hexadecimal notation.

| Base 10 | | | Base 16 | | | Hexadecimal word |
|---|---|---|---|---|---|---|
| 100 | 10 | 1 | 256 | 16 | 1 | |
| | | 0 | 0 | 0 | 0 | 000 |
| | | 1 | 0 | 0 | 1 | 001 |
| .............. | | | | | | |
| | 1 | 2 | 0 | 0 | C | 00C |
| 3 | 4 | 2 | 1 | 5 | 6 | 156 |

Note that now we are using a base larger than ten we have to use letters to represent numbers, i.e.

| Base 10 | 16 | 15 | 14 | 13 | 12 | 11 | 10 | 9 | 8 | 7 | 6 | 5 | 4 | 3 | 2 | 1 | 0 |
|---|---|---|---|---|---|---|---|---|---|---|---|---|---|---|---|---|---|
| Base 16 | 10 | F | E | D | C | B | A | 9 | 8 | 7 | 6 | 5 | 4 | 3 | 2 | 1 | 0 |

Computers may be happy with large streams of binary ones and zeroes but they are totally unmanageable for the human mind. When we think of computer languages such as BASIC we think of short but meaningful phrases. The computer has to compile and assemble this shorthand we call computer language, through various stages of hexadecimal or octal machine code until we arrive at the binary code the computer can understand.

## 5.3   Bandwidth and data rate

Bandwidth is a measure of a communications channel's capacity to carry useful information. The bigger the bandwidth the more information carrying potential that channel has. The unit of bandwidth is hertz,

Hz, or kilohertz, kHz, megahertz, MHz or gigahertz, GHz. Sometimes the bandwidth might be linked to a unit length, such as MHz.km. A common mistake is to refer to length related bandwidth as megahertz *per kilometre*, MHz/km. This is wrong as it implies that the longer the channel the more bandwidth we will have. This is never the case.

### 5.3.1 Data rate

Users of communications systems are not really interested in bandwidth; they are more concerned with getting useful work out of the communications channel such as bits and bytes their computers can use. The units here are bits per second b/s, or kb/s or Mb/s and so on. If we are considering 8-bit words, or bytes, then the units are kilobytes per second, kB/s.

## 5.4 Megabits and megahertz are not the same!

But they are related. The more bandwidth we have, the easier it is to send data down that channel.

Consider the bandwidth as a large highway, as in Fig. 5.4, where the number of traffic lanes represents the bandwidth. The useful work we get out of the highway is the amount of cargo carried by trucks down it. How can we get more cargo down the highway? Well we can add more lanes, this is the same as adding more bandwidth. But we can also make the trucks go faster and make them carry more cargo. This is analogous to the numerous and sophisticated coding techniques that squeeze more data out of restricted bandwidth channels.

### 5.4.1 Measuring bandwidth

As we have seen, bandwidth is a measure of the information carrying capacity of a channel. In engineering it has a very precise definition. The amount of gain or attenuation a signal suffers as it passes through the channel will depend upon the frequency of the signal. All frequencies within a signal will not be attenuated to the same amount

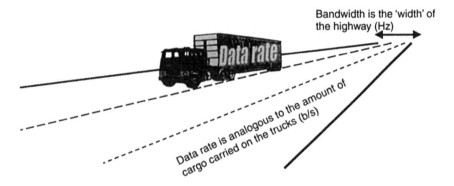

**Fig. 5.4** Bandwidth highway.

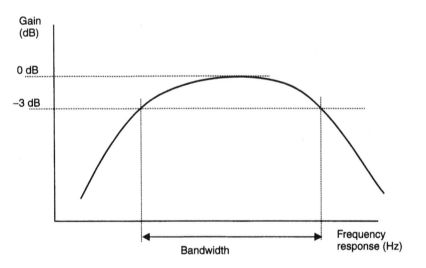

**Fig. 5.5** 3 dB bandwidth of a channel.

within the communications channel. Typically the higher frequencies will be attenuated more, but there will come a point where even the lower frequencies become attenuated. The area in the middle is called the pass-band. Some circuits are especially designed to attenuate or amplify particular blocks of frequencies, in which case they are called filters. The definition of the bandwidth is the frequency band between the frequency points where half of the power has been lost to attenuation, the so-called 3-dB points, because 3 dB represents half power (see Fig. 5.5).

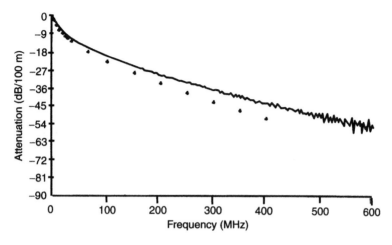

**Fig. 5.6** Attenuation against frequency of a typical category 5 copper cable link.

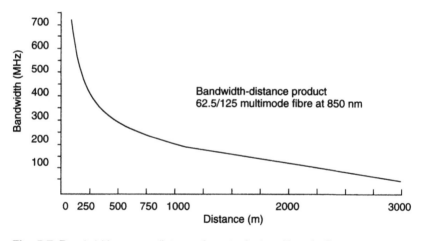

**Fig. 5.7** Bandwidth versus distance for a typical multimode fibre.

The bandwidth definition is very precise but it is often used as an expression with the slightly more cavalier interpretation of the amount of useful data throughput. In structured cabling the word 'bandwidth' is often used to describe the frequency range from zero to the point where the ACR (attenuation to cross talk ratio) goes to zero. This is not strictly correct but does give an idea of the useable frequency range of the cabling system. The 3 dB bandwidth of structured cabling, over 100 m, is more in the region 10–20 MHz, but it would be unlikely to be ever presented as such. Figures 5.6 and 5.7 give typical bandwidth performances for copper and optical cable.

## 5.5   Noise

Noise is the unwanted addition to the desired signal. Noise is the limiting factor in the amount of data that can be transmitted through any channel. Noise can come from anywhere, it is *any* unwanted signal. White noise is energy randomly spread across a wide frequency band. In audio terms we would hear a rushing hiss that seemed to contain all the notes from high to low. We hear white noise when we tune off a station in the VHF band because the radio is giving us an audible interpretation of the white radio noise it is picking up and also creating within its own circuits. In a jet aeroplane the sound of the slipstream rushing over the fuselage gives a white noise sound within the cabin. It is called white noise after white light, which in reality contains all the colours of the visible spectrum (as demonstrated when white light is split into a rainbow by a prism) but which our eyes see as white light.

Noise can also be random impulse noise generated externally. Such sources can include lightning, mobile phones, radars, fluorescent lights, any electrical switching device or even sunspots. Every active electronic device through which the signal passes will add noise, even an amplifier. An analogue signal cannot be regenerated an infinite number of times. Every time the signal passes through the amplifier circuit a small amount of noise is added, which can never be removed. It is like making a copy of a copy of a copy of a VHS videocassette. Every copy will be slightly poorer than the one before. Digital signals can be repeated almost infinitely as long as the process happens before too much noise has been added. In the digital repeater a decision circuit will decide upon whether the input signal is a 'one' or a 'zero' (which it can do if not too much noise has been added) and then reproduce a brand new identical copy with no accumulated noise. Digital systems also have the benefit of error correcting codes that can make up for errors to some extent, but at the sacrifice of some of the system bandwidth.

## 5.5.1   Signal to noise ratio (SNR)

A communications circuit will work when the size of the desired signal is very large compared to the interfering noise. We have a problem when the desired signal starts to become lost amidst the noise. The ratio of the desired signal to the interfering signal is called the signal to noise ratio (SNR) and its unit is the decibel.

$$SNR = 10\log\left(\frac{\text{desired signal power}}{\text{noise power}}\right)$$   [5.1]

Obviously the SNR must be as high as possible to ensure reliable operation of a communications link.

Figure 5.8 shows how a 'clean' digital signal entering a communications medium will emerge at the other end after suffering the effects of attenuation and noise.

Noise in a cabling system comes from three places;

1   External noise. This can be from anything, lightning, mobile phones etc.
2   Within the same cable sheath or bundle. Pickup from adjacent pairs or cables is called crosstalk. As we shall see in chapter 9 there are many different types of crosstalk. Crosstalk within the cable is the largest source of noise within a cabling system.
3   Within the twisted pair, i.e. within the communication channel itself. Return Loss is reflections back down the pair due to impedance mismatch.

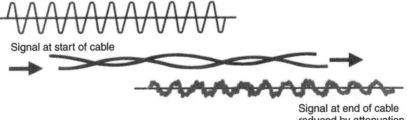

Signal at start of cable

Signal at end of cable reduced by attenuation and with added noise

**Fig. 5.8** Noise and attenuation in a communications channel.

### 5.5.2 Bit error rate (BER)

The owner or user of the communications channel is not so much interested in the SNR as in its practical effects. The effect of noise in a digital communications channel is to degrade the bit error rate (BER). The bit error rate is a measure of how many bits are received in error compared to those transmitted. Ideally if you send a stream of binary 'ones' and' zeroes' down a link they will all arrive in exactly the same order as they were sent. In practice noise will not allow this. Modern LANs such as Ethernet and ATM require a system BER of better than $10^{10}$, i.e. for every ten thousand million bits of data sent, not more than one will be interpreted incorrectly at the receiver. A test device called a bit error rate tester, or BERT, is used to measure this function.

## 5.6   The time domain and frequency domain

A mathematician called Fourier demonstrated how any cyclic train of pulses, of whatever shape, can be made up of sine waves superimposed upon one another. Analysis of these sine waves, as seen in Fig. 5.9, reveals the bandwidth requirements of the pulse train if we want to successfully send it down a communications channel.

If we look at how a signal varies with time, such as a train of digital pulses, we describe this as being in the time domain. We can also consider the frequency spectrum of a signal, i.e. how the amplitude of the signal varies according to frequency. This measurement is called the frequency domain. The two measurements are related. A train of pulses viewed in the time domain, Fig. 5.10, has a frequency spectrum. A piece of mathematics called Fourier transform links the time domain to the frequency domain. The time domain tells us what is happening to data in the channel as a function of time, and can also be used to find out distance related problems with the channel. The frequency domain tells us about the bandwidth requirements of the signal.

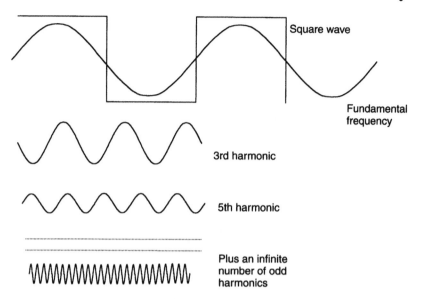

Fig. 5.9 Decomposition of a square wave into sine waves.

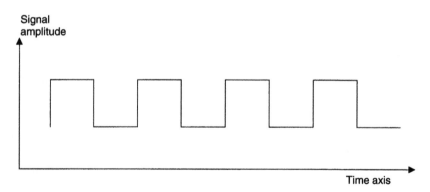

Fig. 5.10 Pulse train in the time domain.

This kind of simple symmetrical pulse train is linked to its frequency domain spectrum by the following Fourier series:

$$f(t) = \frac{4}{\pi}\left(\cos \omega t - \frac{1}{3}\cos \omega t + \frac{1}{5}\cos 5\omega t \cdots\cdots\right)$$   [5.2]

We can see that the pulses are composed of a fundamental frequency ($\omega t$), and an infinite number of odd harmonics. $\omega$ is known as the angular frequency and equals $2\pi f$. To get an exact replica with 90° edges on each corner we would need an infinite bandwidth with all the odd harmonics up to infinity. This is clearly not possible and real life signals on a cable would never look as neat as a textbook diagram.

Again going back to real life situations, nobody sends real data as a continuous stream of 'ones' and 'zeroes', this would only happen if one were sending a clock signal for some form of synchronisation. Real data is in effect a random stream of 'ones' and 'zeroes' which can occur in any order. Simple unipolar NRZ or non return to zero coding is literally a positive voltage for '1' and a negative voltage for a '0'. But the numbers of 'ones' and 'zeroes' occurring next to each other is unpredictable.

There will be many frequency components in a Fourier spectrum of such a random or pseudorandom data stream, but taken together there will be an envelope curve incorporating them all, called the sine $x$ over $x$ envelope, equation 5.3, where $x = \omega_n \tau/2$, and $\omega_n$ is the angular frequency of the '$n$th' term and $\tau$ is the pulse width.

$$\frac{\sin x}{x} \qquad\qquad\qquad [5.3]$$

This function goes to zero at the reciprocal of the shortest pulse width. If we take the LAN and telecommunications protocol called ATM (asynchronous transfer mode) which uses NRZ coding for copper cable transmission at 155 Mb/s, the shortest pulse length would occur when we send a stream of 'ones' and 'zeroes' so the zero crossing points will be at 155 MHz and multiples thereafter, as seen in Fig. 5.11.

With NRZ we see that the frequency spectrum extends to infinity, as it would have to if we wanted to reproduce square waves with precise right-angled corners, i.e. it requires an infinite bandwidth. In reality most of the energy lies within the 100 MHz band for a 155 Mb/s NRZ signal, and a bandwidth of this size would adequately repro-duce the signal but with the addition of jitter, an indeterminate tran-sition point from negative to positive and vice versa, caused by the loss of the higher frequencies. Jitter will accumulate over numerous transits through channels and repeaters and will become a limiting

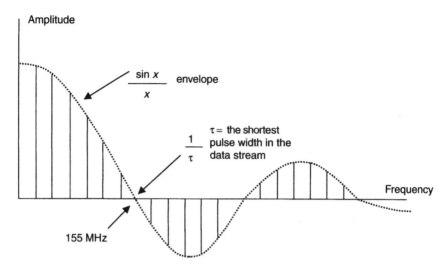

**Fig. 5.11** Sin*x*/*x* function for 155 Mb/s ATM (NRZ random data).

noise factor. We will see from Nyquist's theorem that 77.5 MHz is the absolute minimum bandwidth for a simple 155 Mb/s NRZ signal.

The frequency response of different encoding methods is very different. Coding schemes are often chosen because of the narrower bandwidth they require for transmission. The trade-off however is a greater sensitivity to noise and more complex electronics to execute the encoding and decoding.

Fourier transforms are also used in cable testers. The cable circuit can be analysed in the frequency domain and Fourier transforms applied to convert the data into the time domain. Knowledge of the cable's NVP, which relates to the velocity of the signal down the cable, allows the time axis to be converted to a distance measurement.

## 5.7 Coding schemes

Data can be encoded in basically three ways.

### 5.7.1 Amplitude

At its simplest, a positive voltage represents a digital 'one' and a negative voltage represents a digital 'zero'. But there can be any number of voltage levels. If we had five voltage levels then each state could represent up to five three-digit binary words.

## 5.7.2  Frequency

One frequency can represent a 'one' and another frequency can represent a 'zero'. A more complex system could use many frequencies to represent a number of $n$-digit binary words.

## 5.7.3  Phase

The frequency can remain the same but its phase relative to an initial reference point can change to represent 'ones' and 'zeroes'.

A coding system may use any one or a combination of one or more of the above techniques.

There are many coding schemes in existence and it would take a dedicated book to describe them all. An engineer will choose a coding scheme for a LAN protocol depending upon:

- its efficient use of bandwidth
- its tolerance to noise
- its ease and economy to produce in silicon
- its acceptance by other manufacturers as a 'standard'.

Figure 5.12 shows the effect on system bandwidth when the same data rate is encoded onto several different coding schemes.

Some commonly used schemes are:

| | |
|---|---|
| NRZ | Non-return to zero, or unipolar. Simple to implement with excellent noise rejection but extravagant use of bandwidth. Used on ATM. |
| Manchester | Mixes a clock signal with the original data stream for ease of synchronising the receiver circuit. Used on 10baseT Ethernet. |
| MLT3 | A three voltage level code. Used on Fast Ethernet 100baseTX. |

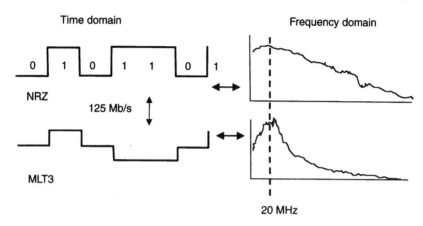

**Fig. 5.12** The effect on bandwidth of different coding schemes.

PAM 5      A five level pulse amplitude modulation code. Used on gigabit Ethernet 1000baseT.

QAM        Quadrature amplitude modulation. Uses many voltage levels and phase differences to produce a constellation of signals. Packs a lot of data into a small bandwidth but subsequently sensitive to noise. Used on xDSL (digital subscriber line) transmission systems.

Some others are:

Differential, differential manchester, duobinary, alternate mark inversion;
PR4, carrierless amplitude phase, trellis;
8B/6T, 2B1Q.

Common local area network coding schemes for copper cabling:

| LAN | Cable media | Coding scheme |
|-----|-------------|---------------|
| 10baseT | 2 Pair category 3 | Manchester |
| 100baseT2 | 2 Pair category 3 | 5 Level PAM |
| 100baseT4 | 4 Pairs category 3 | 8B/6T |
| 100baseTX | 2 Pairs category 5 | MLT3 |
| 1000baseT | 4 Pairs enhanced category 5 | 5 Level PAM |

We can see that some protocols use more than the traditional two pairs i.e. one pair for transmit and another for receive. This is another trick to get yet more out of a cable. A standard four pair cable is after all four communications channels within the same sheath. Gigabit Ethernet, 1000baseT, essentially divides up the 1000 Mb/s data into four 250 Mb/s sections and sends one of the data streams down each of the four pairs of the cable. Logic circuitry at the far end rebuilds the data stream into 1000 Mb/s. Not only does it do this but it simultaneously transmits the send and receive data down the same cable pair at the same time, so achieving full duplex transmission!

Note: Simplex means sending information in one direction only. Half duplex means sending information in both directions, but not at the same time. Full duplex means sending information in both directions at the same time.

## 5.7.4   Nyquist theorem

A mathematician called Nyquist demonstrated the minimum sampling frequency needed to reproduce a signal. An analogue signal is converted to a digital signal by an analogue-to-digital-converter (ADC), and back again by a digital to analogue converter, or DAC. The analogue signal has to be sampled at various points in time and the instantaneous value of signal voltage recorded at that moment in time is converted into a binary number. We can see that the 'timing gate' must be very short compared to the rate of change of the sampled signal. To reproduce the analogue signal it would seem obvious that we sample and digitise it as often as possible, but we know bandwidth is expensive and must be used as efficiently as possible.

Nyquist shows that a signal must be sampled at least twice as often as the highest frequency contained in the measured signal.

Take a telephone channel, which needs 3.3 kHz of bandwidth for an analogue voice signal. To digitise it we need a sampling rate of at least 6.6 kHz. In practice 8 kHz is more likely to be used. We thus have 8000 samples of information every second. If an eight-bit word is used (giving 256 different levels) to represent each sample then the digital data rate becomes $8000 \times 8 = 64$ kb/s. A video signal needs an 8 MHz bandwidth. If sampled at 16 MHz and 8-bit encoding is used then the data rate for that signal becomes 128 Mb/s. From earlier theory we know that such a signal needs 64 MHz of channel bandwidth for transmission, eight times more than what we started with! Some might ask, what is the point of digitisation?

In practice, coding schemes are available which compress the required bandwidth by taking out redundant data. For example a television picture may contain 50% of blue sky. Instead of coding every pixel on the screen a code can be used which says that, 'the following number of bytes are all the same, so I'm only going to send one of them once'. A good quality television picture can be compressed to 8 Mb/s.

Digitisation makes regeneration of the signal much easier without adding noise, and facilitates switching and digital storage.

As we noted, an 8-bit word gives up to 256 different possible values to express the original value of the sampled analogue signal. But an analogue signal is infinitely variable between its minimum and maximum points. The difference between the discrete value of the digital word and what the value was in reality is called quantization noise. We must sample at least twice the maximum frequency component of the analogue signal but from then on we can only improve noise by using larger binary words (see Table 5.1). CD quality music may use 12 or 16 bit words, but this is at the expense of

Table 5.1 Quantization SNR improvement with larger binary words for each symbol

| Signal to noise ratio (dB) | No. of levels | Binary word length |
| --- | --- | --- |
| 11 | 2 | 1 |
| 17 | 4 | 2 |
| 23 | 8 | 3 |
| 29 | 16 | 4 |
| 35 | 32 | 5 |
| 41 | 64 | 6 |
| 47 | 128 | 7 |
| 53 | 256 | 8 |

consuming more bandwidth in transmission or memory space in a recording.

### 5.7.5  Shannon channel capacity

In 1949, CE Shannon demonstrated that if one knew two factors about a channel, namely the bandwidth and the signal to noise ratio, then it was possible to predict the maximum information carrying capacity of that channel.

$$C = W \log_2\left(1 + \left[\frac{S}{N}\right]\right) \tag{5.4}$$

Where  $C$ = channel capacity, in bits per second
  $W$ = channel bandwidth, in hertz
  $S/N$ = channel signal to noise ratio, in decibels.

In a NEXT dominated cable channel the following can approximate the channel capacity:

$$C = F \log_2(Af - 1) \tag{5.5}$$

Where $F$ is the frequency at which the ACR goes to zero

> $Af$ is the actual channel attenuation at that frequency and assuming the attenuation is much greater than 1.
> (Although this equation makes it appear that more attenuation will give you more channel capacity, this is a special case where NEXT = attenuation, i.e. ACR = 0. More attenuation would move the attenuation line up so that the frequency at which ACR = 0 occurs ($F$) will also reduce.)

For a category 5/class D (100 m) channel (presuming ACR goes to zero at 100 MHz) gives a capacity of 437 Mb/s per pair. If we consider that a category 5 cable in fact consists of four pairs, each of which is a communications channel in its own right, then the cable capacity becomes 1.75 Gb/s over 100 m.

The proposed category 6/class E gives figures of 975 Mb/s per pair and 3.9 Gb/s per cable, presuming ACR goes to 0 at 200 MHz.

This is of course a theoretical maximum. Even though NEXT dominates as the principle cause of noise we would still have the effects of FEXT, return loss and external noise to contend with. We have had the mathematics to calculate channel capacity since 1949, but until the advent of cheap processing power in the form of microprocessors, the idea of encoding and signal processing for a mass market was just not practical. Gigabit Ethernet transmits 1000 Mb/s over enhanced category 5 cable (100 m channel) and it is proposed to send 2.5 GbE (i.e. 2.5 Gb/s) over category 6. These figures are likely to represent maximum practical values for these classes of cables.

# 5.8  Multiplexing

Multiplexing is a way of sharing a number of transmission circuits on one communications channel. It can be achieved in one of two ways, time division multiplexing (TDM) and frequency division multiplexing (FDM). Optical fibre also has the capacity to accept wavelength division multiplexing.

## 5.8.1  Time division multiplexing (TDM)

It is easier to visualise this concept by imagining two mechanical rotary switches; one attached to each end of the line. Various circuits are connected to the other side of the rotary switches that are circulating in a synchronised pattern to each other, as seen in Fig. 5.13. As the switches make contact from one circuit to another a hard-wired connection is made for each one from one end to the other, and each circuit has its own turn at being connected.

The switches must rotate significantly faster than the data is changing on the input line or else data will be lost. The aggregate data rate on the channel will be the sum of all the different data rates on each input line. A TDM is an electronic version of this rotating electro-mechanical switch concept. In practice it is unlikely that every input line will be sending data permanently at its maximum rate. So it is possible to share a channel amongst a number of input lines whose

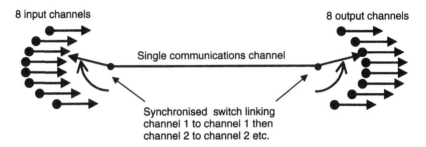

**Fig. 5.13** Time division multiplexing.

aggregate data rate exceeds the channel capacity, on the assumption that the maximum input data rate will be rarely reached. Such a device is called a statistical multiplexer, and this is a much more efficient device for sharing telecommunications links where bandwidth is expensive. If data rates do start to exceed the channel capacity the statistical multiplexer is equipped with a buffer memory to hold some of the data until channel capacity becomes available. If the buffer memory becomes full then the multiplexer must have the ability to communicate with the sending devices to tell them to wait before sending any more data.

## 5.8.2   Frequency division multiplexing (FDM)

Apart from the time domain the channel can also be shared in the frequency domain. Filters can be used to divide the channel up into a number of frequency bands, as in Fig. 5.14, and for example a 10 MHz channel could be divided up into ten 1 MHz bands and the data or other information could be sent within each band. FDM is particularly suitable for analogue transmission such as a number of analogue video signals sharing one common channel.

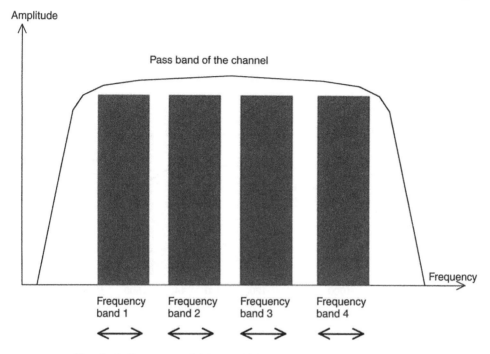

Amplitude

Pass band of the channel

Frequency

Frequency band 1    Frequency band 2    Frequency band 3    Frequency band 4

←→    ←→    ←→    ←→

**Fig. 5.14** Frequency division multiplexing.

### 5.8.3    Wavelength division multiplexing (WDM)

Optical fibre transmission is usually considered according to the wavelength of light used rather than its frequency (frequency = velocity of the signal divided by the wavelength). The useable bandwidth of an optical fibre can be greatly increased by WDM. With sufficiently accurate optical filters the huge capacity of singlemode fibre, nominally at 1310 nm, can also be obtained at 1316 nm for example, and so on, as in Fig. 5.15. The term dense wavelength division multiplexing (DWDM) is used when the channel spacing is very close indeed.

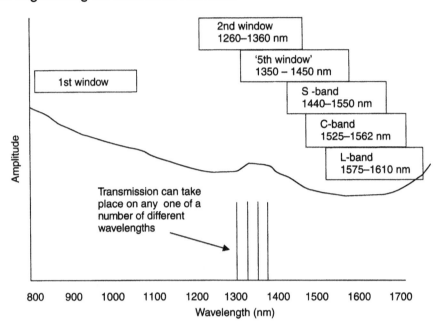

**Fig. 5.15** Wavelength division multiplexing.

# 5.9    Synchronous and asynchronous transmission

In communications the individual 'ones' and 'zeroes' that make up the useful data are usually divided up into some form of packetised structure. In asynchronous transmission each packet of data is a self-contained entity. It has bits at the beginning to show the receiver that this is the start of the packet, a payload of the useful data in the middle and another number of bits at the end to show where the end of that packet is. It may also contain destination and source information and error correcting codes of varying complexity. Asynchronous data is very easy to route around the world with various payload sizes and the speed of operation can be very easily scaled up or down. The disadvantage is that the extraneous bits surrounding the payload represent a significant overhead and valuable channel bandwidth is soaked up with this non-payload data. Asynchronous transmission does not make the best use of the available channel bandwidth but this may sometimes be acceptable due to asynchronous transmission's other benefits.

Synchronous transmission requires synchronisation between transmitter and receiver so that the receiver knows when to sample the incoming data and what manner that part of the data stream represents, e.g. addressing data. Far less overhead is required to delineate discrete packets of data. The transmission network may require an overall system clock to keep everybody in time or else a clock signal may be mixed up with the data to enable local synchronisation at the receiver.

## 5.10    Error correcting codes

Codes, simple or complex, may be added to the data packets and which have some numerical relationship to the payload data. The receiver can make a calculation on the payload data and the result compared to the original error code information compiled by the transmitter. If the result is the same then the chance that the data has been received without error is very high. If the answer is different then the data has probably been corrupted and the receiver may ask the transmitter to send that packet of data again. Error correcting codes can mask occasional impulse noise on the communications channel as a corrupted data packet is simply retransmitted. A serious noise problem will first manifest itself as an apparent slowing down of the network, or long response times at each affected terminal as the data is repeatedly retransmitted. But after a point the error correcting codes will be unable to cope and the system will collapse. Communications links without error correcting codes will see noise as corrupted data on the terminal screen and is often associated with mysterious logging on or off to the mainframe or server.

## 5.11    Digital subscriber lines (xDSL)

xDSL is a family of technologies developed to send high-speed data over copper access lines. It is primarily a telecommunications technology and application but interaction with LAN and premises structured cabling systems is inevitable. Access to the telecommunications environment, such as the internet, over twisted pair telephone lines is currently achieved by using analogue modems with a maximum speed of 56.6 kb/s or by integrated services digital network (ISDN) which offers a data rate of 64 kb/s. A modem (modulator/demodulator) encodes data to fit within the standard 3.3 kHz analogue transmission band of telephone lines. DSL technology aims to significantly improve on that transmission performance over twisted pair cables.

xDSL can be symmetric or asymmetric. Symmetric means that the data rate in both directions is the same. Asymmetric means that the data rate from the exchange to the subscriber is much larger than the rate from the subscriber back to the exchange. The physical separation or distance from the transmitter to the receiver determines the maximum achievable data rate.

HDSL        High speed digital subscriber line. Offers up to 2 Mb/s in each direction (i.e. symmetric). The first 1993 specification called for three pair operation, though the 1996 version used two pairs. There is currently a draft ANSI specification for a one pair version. This will probably only be used for the American T1 data rate (1.5 Mb/s) not the European E1 data rate (2 Mb/s). HDSL uses a band from DC to 748 kHz and uses 2B1Q or CAP encoding.

ADSL        Asymmetric digital subscriber line. Offers 6–8 Mb/s downstream and 640 kb/s–1 Mb/s upstream. There is a standard, ANSI T1.413, which specifies discrete multi tone (DMT) encoding but there is also a non-standard scheme using carrierless amplitude phase (CAP) and quadrature amplitude modulation (QAM). ADSL uses a band of 25 kHz–1.1 MHz.

ADSL-Lite   ADSL-Lite is a 'stripped' down version of ADSL to give a cheaper, user-installable high speed delivery system (1.5 Mb/s) primarily for internet use. The ITU is working on a standard (G.992.2) to define this technology.

VDSL        Very high speed digital subscriber line. This emerging technology pushes data transmission over simple copper telephone lines to the limits. It will be able to offer a symmetric service of 26 Mb/s and an asymmetric service of 2 and 52 Mb/s, but over relatively short distances of around 300 m for the higher speeds. VDSL uses a combination of DMT and QAM and requires a band of 300 kHz–10 MHz.

# 6

# Local area networking and associated cabling

## 6.1  Introduction

A LAN is a collection of hardware and software components that allows and facilitates communications between shared networking devices. These devices can be personal computers (PCs), dumb terminals, printers, storage devices and access devices to the outside world such as routers.

The 'local' in LAN usually means a privately owned site of less than four kilometres in diameter, but the definition of where LANs end and telecommunication systems start is becoming more and more obscure. Where LANs do need to connect to remote sites they have to interface with a telecommunication system and these are then collectively called wide area networks (WANs), or even metropolitan area networks (MANs), if the majority of connections remain within the same city.

The driving force in LAN is the growth in power of microprocessors and the corresponding growth in size of the associated applications software packages that can run on them. More computing power and cheaper memory chips leads users to run more bandwidth-hungry applications such as anything involving video or pictures. The last few years of the 1990s saw the change in microprocessor products from the Intel 486 to the Pentium, Pentium II and

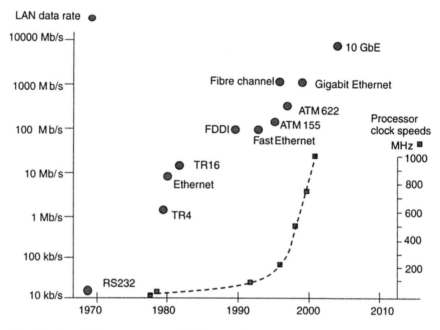

**Fig. 6.1** Growth in processor and LAN speed.

ending the decade with the Intel Pentium III running at 733 MHz and the AMD Athlon running at 750 MHz. 2001 will see the introduction of 64 bit computing, compared with 32 bit, which will give another enormous boost to chip processing power.

To communicate with the outside world the computer needs a high-speed communications link concomitant with the speed and processing power of the microprocessor. The LAN must therefore keep pace with the demands of the personal computer and the structured cabling system must keep ahead of the demands of the LAN. We can see in Fig. 6.1 that 20 years of experience has shown us that LAN speeds have grown at a rate one hundred times, or 10 000% per decade!

## 6.1.1   The components of a LAN

LANs connect computers, peripherals and storage devices together and need the appropriate hardware and software to achieve this. A computer may be defined as a central processor, a range of memory devices, a visual method of reading data from the computer, i.e. a screen, a method of inputting data to the computer, i.e. the keyboard and mouse, power supplies, an input/output device for communicating to the outside world, and of course software. The software may be defined in three layers:

- Application software, such as word processing.
- An operating system to control the programs and their interaction with the outside world, e.g. Windows.
- A network operating system (NOS), to control the behaviour of the computer on the LAN, such as Novell Netware and Windows NT.

The computer makes its connection to the LAN via a network interface card (NIC), which plugs into the backplane of the PC and presents an industry standard connector to connect into the LAN media. There are NICs for all of the popular LANs such as Ethernet and Token Ring. Obviously they all have to be of the same type to communicate with each other.

The LAN needs some form of communication media and this may be based on cable or wireless technology. The cable can be copper or optical fibre and the wireless method can employ radio or infrared technology. The computer should be indifferent to the communications media as the standardised software and hardware interfaces should take care of all the details. Other devices can be connected to the LAN such as printers, shared memory resources or even fax machines.

The LAN can be peer-to-peer where all users have an equal say in what happens on the LAN but this would soon be chaotic on a larger site and it is usual to have network supervisory stations and servers which act as central memory devices and as arbitrators of varying security levels and who has access to what. We should also remember the people; users, supervisors and network managers who use and make the whole enterprise work.

# 6.2   The rise of structured cabling and local area networks

There are many different types of LAN but Ethernet is the dominant technology today. The purpose and whole philosophy of structured cabling is to provide an applications-independent communications medium. This gives the end-user complete choice and control over what protocols and applications he/she can run. When computers were first generally deployed it was the practice for manufacturers to design their own cables specifically for a certain range of computers. There were a few standards such as RS 232 but more often companies such as IBM would specify a certain cable for every computer system in their product range. Some examples are given in Table 6.1.

The network architecture was either point-to-point or daisy-chain. Point-to-point means that a cable is deployed from the central processor directly to, and dedicated solely to, the peripheral.

Daisy-chain means a cable running from the central processor to the first peripheral and then linking on to the next peripheral and so on. Some form of addressing or polling protocol is required to allow each peripheral to know when it is being addressed and for the central processor to be able to recognise which peripheral has communicated with it.

The disadvantages of this dedicated form of wiring are:

- It is dedicated and proprietary to a particular manufacture and/or computer system.

Table 6.1  Computer systems and their original cable media

| Computer system | Original cable medium |
| --- | --- |
| IBM 3270 | 93 Ω coax |
| IBM AS400 | 105 Ω twinax |
| IBM RS6000 | RS 232 multicore |
| Wang VS | 75 Ω twinax |

- Moves and changes are costly and time consuming as provision for future expansion is rarely implemented.

A building of the 1970s or 1980s may have had many different types of cables for different applications, such as:

- Computer cables as listed in Table 6.1.
- 2, 3 or 4 pair cables for telephones and PBX extensions.
- 25, 50, 100 and 300 pair cables for telephony backbone cabling.
- 75 Ω coax for CCTV/security video.
- Electronic point of sale (EPOS) cabling.
- Various and numerous cables for security, access control, heating and ventilation control (HVAC), public address system, environmental monitoring, smoke and fire detection and fire alarms.
- 'Dealer desk' dedicated video cabling such as at Bloomberg and the RGB triple coax for Reuters.

Structured cabling can in theory cope with all of these requirements on one common set of cables, with subsequent savings in hardware, installation, management and changes.

Today all data applications run on the structured cabling system and most, but not all, PBX telephone systems. About half of all video applications now run on the structured cabling system except for broadband video distribution (e.g. CATV) whose 550 MHz bandwidth requirement is still too demanding for category 5 twisted pair cabling.

The building control systems, generally grouped under the heading 'intelligent building', have recently started to utilise the benefits of structured cabling but at a rate slower than most anticipated. Some fire detection, alarm and emergency lighting circuits are required to use fire survival cable such as mineral insulated cable.

IBM is generally credited with the introduction of structured cabling around 1984. IBM proposed that a set of data cables should follow the same philosophy as telephone cabling, i.e. a common set of flood-wired cables that terminate at every position where a user sits or even where somebody may sit in the future. All the cables would terminate in a common connector at the user end and at a set of patchpanels at the administration end. By means of the patch panel any user could be connected to any application. The initial investment may be more but the lower cost of moves and changes would give the structured cabling system a payback period of between three and five years.

The IBM Cabling System (sometimes referred to as ICS) utilised a twin pair screened cable of $150\,\Omega$ characteristic impedance. A brand new hermaphroditic connector was designed to complement the cable, and the application firmly in the designers' minds at the time was the IBM Token Ring LAN.

Token Ring was launched as a 4 Mb/s LAN though 16 Mb/s soon followed. The IBM Cabling System however has a frequency performance of at least 300 MHz, and what with the screening elements and large connector it soon became apparent that it was relatively over-engineered for the job it had to do. The IBM Cabling System has since matured into the advanced connectivity system (ACS) which offers standards-based, $100\,\Omega$ connectivity, yet in true IBM tradition still offers the user the option of a completely proprietary connector.

By 1988 a part of AT&T, later trading as Lucent Technologies, advanced the idea of using simple four pair, $100\,\Omega$, unshielded twisted pair cabling direct from the American telephone network. 110 cross connects and the 8-pin connector we now refer to as the RJ45 were pressed into service and used to demonstrate at least 1 Mb/s operation but at a much lower cost than the IBM Cabling System. Like ICS, AT&T's Premises Distribution System could link up 'legacy' computer and cabling systems by the use of a balun, a device that converts from one cable type, and impedance, to another. The AT&T system formed the basis of the $100\,\Omega$, four pair cabling system still in use today for generic premises cabling systems.

By 1990, many other cable and connector manufacturers were offering look-alike cabling systems all claiming widely differing performance. To offer some guidance to customers, and suppliers, the American distribution company Anixter introduced a grading system for the cable known as 'levels' with level 3 being the highest.

In 1990 the market started to open up with the introduction of the 10baseT Ethernet standard. Although the first forms of Ethernet used coax, 10baseT was designed to work on good quality telephone cabling, hence the 'T'. 10baseT and structured cabling went hand-in-hand and the market accepted the whole concept and benefits of structured cabling within three years.

By the early 1990s the various bodies that write standards started to catch up with events, specifically the TIA, the Telecommunications Industry Association of the US. The TIA changed the word 'level' to 'category' and category 3 was born. The TIA, in association with the EIA and ANSI (see chapter 15 for details of the standards) published TSB 36 for cable and TSB 40 for connecting hardware.

Category 3 described a cabling system with a 16 MHz bandwidth and with 10baseT firmly in mind. When IBM introduced their 16 Mb/s Token Ring LAN they pointed out that the cabling system needed a bandwidth of at least 20 MHz. Category 4 was hurriedly rushed out offering a 20 MHz bandwidth, but it had a life measured in months as the designers already had their eye on 100 Mb/s Ethernet. Hence 100 MHz category 5 was born and published as a standard in 1995, closely followed by the International ISO 11801 and European CENELEC EN 50173. The cabling systems described in 1995 lasted for the rest of the decade but for the new millennium a range of advancements in structured cabling and LAN technology, such as category 6 and 7 and gigabit Ethernet, are now firmly with us.

# 6.3   LAN architecture and the open systems interconnection model

The International Standards Organisation (ISO), based in Geneva, developed a seven-layer reference model to describe the functionality and relative position of different elements within a computer networking system. It is called the Open Systems Interconnection Model, or OSI model. First developed in 1978, the OSI model allows designers and system engineers to concentrate on the functions required within a particular level of operation and have an agreed set of interface methods to the layer above and below. For example some Ethernet switches are described as 'layer 3' or 'level 3', meaning they communicate with other devices using protocols agreed for layer 3 of the OSI model. The OSI model is shown in Fig. 6.2.

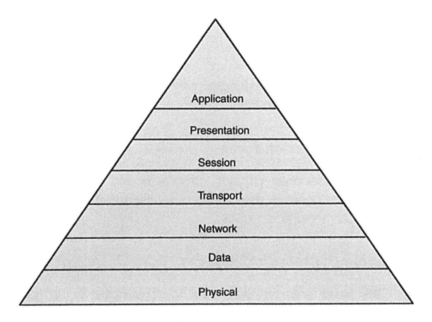

**Fig. 6.2** OSI seven layer model.

## Layer 1, physical layer

The physical layer is of most relevance to those involved in the structured cabling end of the LAN business. But layer 1 is not just electrical and optical passive hardware as it is also concerned with the electrical and optical interfaces themselves. LAN repeaters work at this level.

## Layer 2, data link layer

This layer is responsible for providing error-free transmission of packets between networks. LAN bridges work at this level.

### Layer 3, network layer

The network layer routes data from one network device to another possibly using one or more different routes. Routers can operate between different layer 1 and 2 protocols, e.g. Ethernet and Token Ring.

### Layer 4, transport layer

Layer 4 is responsible for providing a quality of service for data transfer between two devices.

### Layer 5, session layer

The session layer controls, organises and synchronises the data exchange between devices communicating with each other, such as who can send and receive.

### Layer 6, presentation layer

This layer ensures that data is meaningful to all communicating devices such as specifications of character code and data compression.

### Layer 7, application layer

The application layer allows dissimilar application processes to exchange information.

There is sometimes confusion where higher level protocols sit relative to cabling and LAN technology. For example, is it relevant to ask, 'can your cabling system support IP (internet protocol)?' IP is a layer 3 and 4 protocol that provides packet segmentation and re-assembly and has specific addressing conventions (the ubiquitous IP address), it therefore sits 'above' the data link and physical layers which have the more familiar names such as Ethernet and Token Ring. So the answer is yes but only after the IP message is packaged within a layer 1 and 2 protocol.

## LAN equipment

### Network interface card (NIC)

The NIC is the printed circuit board that plugs into the PC bus or expansion slot. It takes raw data from the PC microprocessor and formats it into a standards based protocol with a physical electrical or optical connector. It will present an RJ45 socket for copper cabling and an ST or SC optical connector (although this may change) for optical fibre connection. Many Ethernet cards are known as 10/100 auto-negotiating. This means they will communicate with the hub or switch to see how fast that device can work and then both will work at the highest common speed.

### Repeaters

Repeaters act as regenerators of the signal to allow link extensions.

### Bridges

There are different types of bridges offering different levels of complexity. At their simplest they connect two different LANs of the same architecture. Bridges can also act as repeaters but they offer more functionality. They can speed up the operation of a LAN by segmenting it so those local devices that mostly communicate with each other are not tying up the whole LAN. Bridges read the destination of the incoming packet and can teach themselves which devices are local and which are not. Local traffic is rebroadcast to all the local devices while packets with remote addresses are sent on to the next segment.

### Hubs and switches

Bridges and repeaters hail from the days when LANs were bus oriented, i.e. everybody tapped into a long cable. Hubs and switches collapse the backbone cable into one box so that their logical interconnection mirrors the physical manifestation of structured cabling. A hub shares out the available bandwidth, e.g. 10 Mb/s for Ethernet 10baseT, amongst all those connected to it, just as Ethernet coaxial cable would have done. This is low cost but can end-up delivering very low speeds to the user and for many busy users this would nowadays lead to unacceptable network delays. Stackable hubs offer even lower cost per port but with even more users contending for the limited 10 Mb/s bandwidth. Switches are intelligent hubs with basic routing capabilities. Switches can read the target address of the packet and decide if it is local or needs to be dispatched to the next segment. Switches offer full 10 Mb/s capability to all users connected to the switch and it is usual for the switch to connect into the backbone network (the 'uplink') via a 100 Mb/s link. The next generation will of course offer 100 Mb/s to each user with a 1000 Mb/s uplink requiring 10 000 Mb/s on the campus backbone. Some switches are fairly simple to operate but require local management. More sophisticated switches can be remotely managed by a protocol known as Simple Network Management Protocol, SNMP.

## *Routers*

Routers are more intelligent layer 3 devices that can connect between disparate LANs and also manage connection out into the telecommunications environment of MANs and WANs. Routers can make intelligent decisions about the optimum route for a message to take depending on such factors as cost and reliability. Routers can be software based, which are slow, or hardware based ('wire-speed') which are much faster.

## *Gateways*

The term gateway applies to routers operating at all the OSI layers from 4 to 7. Higher level protocol conversion such as TCP/IP, SNA etc can be effected within a gateway. The gateway is more likely to consist of software operating on a mid-range computer rather than as a standalone box like a hub or switch.

## LAN architecture

The components of a LAN can be laid out in different ways. This is sometimes called the architecture or even topology or topography. The three main methods are bus, ring and star.

### *Bus*

Bus architecture can be considered as a long cable which everybody can tap into for access to the shared resources connected to that bus. See Fig. 6.3. The classic example of bus architecture is the original form of Ethernet, whereby a long coaxial cable was run between users and people could tap into it as appropriate. For a bus system to be successful there has to be some agreed method of contention so that more than one user isn't transmitting at the same time, and an addressing system so each user can recognise messages intended for their station.

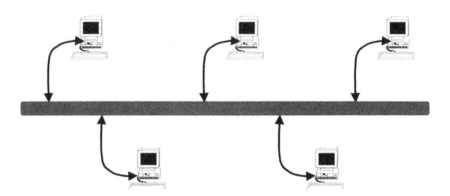

**Fig. 6.3** Bus architecture.

### *Ring*

Ring architecture consists of a cable going from one user to another and then onto another and so on until the first station is reconnected once again. See Fig. 6.4. Token Ring is a logical ring

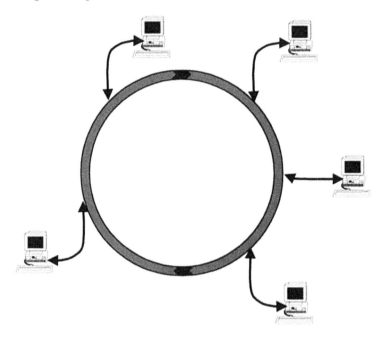

**Fig. 6.4** Ring architecture.

structure. A 'token' is passed from one user onto another and a station is only allowed to transmit if it is in possession of the token. As the token passes by then the receiving station can check to see if any message packets accompanying it contain that station's address.

## *Star*

Star architecture means every user is wired back to some central point which then sorts out contention and addressing issues. ATM can be considered as a star-type architecture. It should be noted that structured cabling is all star wired. Every user is directly cabled back to a central patch panel. This makes structured cabling suitable for all kinds of modern LAN. Ethernet, for example, is still a bus-based system but the bus is now collapsed inside the centralised hub or switch resident in the telecommunications closet. One can imagine

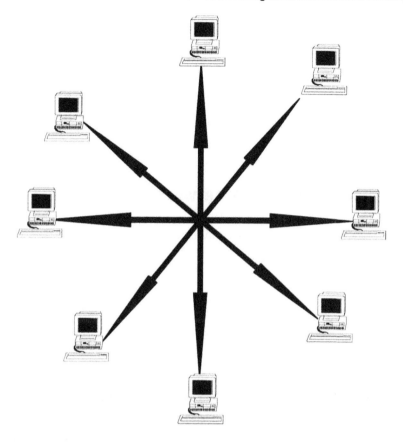

**Fig. 6.5** Star architecture.

the 500 m of coax has been shrunk inside the hub or switch and the AUI drop cables have been replaced by the 90 m links of category 5 cable. Similarly for Token Ring we have the ring condensed inside the multi-station-access-unit, and the 90 m of structured cable now links the station into that collapsed ring (see Fig. 6.5).

## *Baseband and broadband*

The 'base' in 10baseT etc. means that the signal is baseband, i.e. it is not modulated onto a higher frequency carrier. 10baseT occupies a frequency spectrum of approximately 0–10 MHz.

Broadband signals are modulated onto a higher frequency carrier and are intended for use on coaxial cable. The rise of twisted pair structured cabling and the availability of high speed LANs intended for that medium have suppressed the need for and popularity of broadband LAN. A broadband LAN intended for factory automation (manufacturers' automation protocol, MAP) has been standardised as IEEE 802.4.

# 6.4   Types of local area network

The following local area networks make up most of the market:

- Ethernet.
- Token Ring.
- ATM.
- FDDI.

### Ethernet

Ethernet is by far the market leader in the world of LANs, with between 70 and 80% of market share. It has achieved this position by means of cost, breadth of suppliers and product, easy upgrade path and simple implementation rules.

Ethernet uses a protocol called carrier sense, multiple access, collision detect (CSMA/CD). This means that there is a common bus which any user can access (multiple access). The transmitting station must first 'listen' to the network cable to see if anybody else is currently transmitting (carrier sense) and will back off for a random amount of time (collision detect) if it does detect another signal before it tries again. The Ethernet protocol is very successful but performance drops off rapidly under heavy load conditions.

The first Ethernet system was called 10base5. The '10' signifies the speed, i.e. 10 Mb/s, the 'base' signifies that it is a baseband signal, i.e. from 0 Hz to 10 MHz, and the '5' signifies a 500 m range. Ethernet coax is a large, usually yellow, coaxial cable of $50\,\Omega$ characteristic impedance. An Ethernet LAN system could be composed of five segments that could each be up to 500 m long and joined by repeaters. There could be up to 100 taps into this cable which could support 1024 users. All the users were in effect sharing the same 10 Mb/s bandwidth and so response times would be very poor for a network heavily loaded with traffic. The tap from the coax cable would connect back into the transceiver card on the PC via an 'AUI' cable. Every device connected had its own unique Ethernet address so a receiving station could know which messages were intended for it.

Ethernet segments could be bridged together using an Ethernet bridge and segments could be extended using a repeater.

A lower cost version using a daisy-chain coax RG58 style of cable was introduced and called 10base2, formally known as IEEE 802.3a but also known as 'Thinnet', or 'Cheapernet'. This version of Ethernet was cheaper to implement but supported shorter distances, i.e. five 185 m segments with up to 30 users.

A version called 10baseT (IEEE 802.3i) was introduced in 1990 designed for $100 \Omega$ telephone cable. It utilised a star-wired topology that made an ideal match for structured cabling. The bus was in effect compressed into the Ethernet 10baseT hub, with all users once again sharing the same available 10 Mb/s bandwidth. An advancement on this was the introduction of the Ethernet switch. The switch gives each user connected to it a dedicated 10 Mb/s link thus greatly increasing the throughput for every user.

A fibre optic link was introduced for longer distances and particularly for joining separate buildings on a campus. It is known as 10baseF (IEEE 802.3j).

Fast Ethernet is the name given to a range of 100 Mb/s Ethernet protocols, of which there are four for copper, one for optical fibre and another on its way for fibre. Three of the fast Ethernet protocols are designed to work on category 3 cable, principally because of the large installed base of category 3 cable in the USA from the early 1990s onwards. They are known as 100baseT2, 100baseT4 and 100 VG-AnyLAN.

The latter uses a different kind of protocol called demand access priority and was developed by a different committee, IEEE 802.12. All three types are relatively expensive because sophisticated coding techniques have to be used to overcome the shortcomings of the 16 MHz, category 3 cable.

The optical version of Fast Ethernet is called 100baseFX. It uses the more expensive 1300 nm components and so a cheaper version called 100baseSX is being introduced which takes advantage of the much lower cost 850 nm VCSEL (laser).

As a general rule the backbone LAN technology must be at least ten times faster than the *to-the-desk* technology. Hence Fast (100 Mb/s) Ethernet was required for the backbone after 10baseT switching became commonplace. But with users migrating to 100 Mb/s to the desk then logically 1000 Mb/s, or 1 Gb/s is required in the backbone. If this does not happen then a bottleneck will be created and the network speed will be reduced to less than 10 Mb/s. Hence the following gigabit Ethernet protocols have been introduced:

- 1000baseT      IEEE 802.3ab.
- 1000baseCX     IEEE 802.3z.
- 1000baseSX     IEEE 802.3z.
- 1000baseLX     IEEE 802.3z.

1000baseT is designed to work on enhanced category 5 cabling (cat5e) for up to 100 m. 1000baseCX is screened cable which can be used for equipment cables up to 25 m, and is thus not considered part of the structured cabling system.

1000baseSX is a short wavelength (850 nm) fibre optic system and 1000baseLX is a 1300 nm optical system. 1000baseSX can use 50/125 or 62.5/125 multimode fibre and 1000baseLX can use multimode or singlemode fibre. The transmission performances of the copper versions of Ethernet are summarised in Table 6.2. Note that many protocols can transmit further than 100 m but the LAN standards have been reconciled with the structured cabling standards to ensure design compliance for any combination of protocols.

An example layout (topology) of an Ethernet system is shown in Fig. 6.6. Optical fibre versions of Ethernet are 10baseF, 100baseFX, 1000baseSX and 1000baseLX. A 100 Mb/s, 850 nm version is planned to be called 100baseSX. 10baseF and 100baseFX (full duplex) can transmit up to 2000 m over multimode fibre. The

Table 6.2 Ethernet transmission performance summary (copper)

| Protocol | Coding method | Number of pairs used | Cable type required | Transmission distance (m) |
|---|---|---|---|---|
| 10baseT | Manchester | 2 | Cat 3 | 100 |
| 100baseT2 | 5 level PAM | 2 | Cat 3 | 100 |
| 100baseT4 | 8B/6T | 4 | Cat 3 | 100 |
| 100baseTX | MLT3 | 2 | Cat 5 | 100 |
| 100VG-anyLAN | 5B/6B | 4 | Cat 3 | 200 |
| 1000baseT | 5 level PAM | 4 | Cat 5e | 100 |

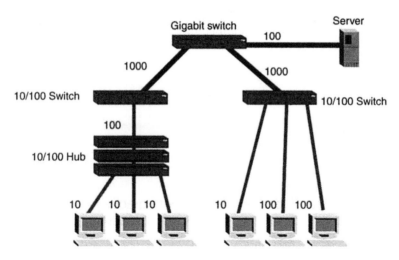

**Fig. 6.6** Typical Ethernet LAN topology.

half-duplex form of 100baseFX is limited to 412m. There is a complicated set of rules governing the distances allowed when different types of Fast Ethernet segments are linked and this is shown in Table 6.3.

Gigabit Ethernet over optical fibre has yet another set of rules. Unlike most optical data transmission systems, gigabit Ethernet is bandwidth limited. Most other systems are attenuation limited. Three different types of optical fibre are allowed; 50/125, 62.5/125 and

Table 6.3 Ethernet collision domain diameters per IEEE standard 802.3u

| Model | All copper (m) | All fibre (m) | T4 and Fibre (m) | TX and Fibre (m) |
|---|---|---|---|---|
| DTE to DTE | 100 | 412 | N/A | N/A |
| 1 Class 1 repeater | 200 | 272 | 231 | 260.8* |
| 1 Class 2 repeater | 200 | 320 | 304 | 308.8* |
| 2 Class 2 repeaters | 205 | 228 | 236.3 | 216.2† |

*Assumes 100 m of copper and one fibre link.
†Assumes 105 m of copper and one fibre link.

Table 6.4 Optical gigabit Ethernet requirements as per IEEE 802.3z

| Fibre type | Fibre bandwidth (MHz.km) | | Transmission distance at 850 nm (m) | Transmission distance at 1300 nm |
|---|---|---|---|---|
| | At 850 nm | At 1300 nm | | |
| 62.5/125 | 160 | 500 | 220 | 550 |
| 62.5/125 | 200 | 500 | 275 | 550 |
| 50/125 | 400 | 400 | 500 | 550 |
| 50/125 | 500 | 500 | 550 | 550 |
| Singlemode | | | | 5000 |

singlemode. Two different bandwidth grades are allowed within each of the two multimode fibres. The quality of the fibre is determined by its available bandwidth, and Table 6.4 demonstrates the link lengths possible with the different fibre types.

Ethernet continues to evolve and now gigabit Ethernet is finalised the IEEE committee has started work on the next generation, 10 gigabit Ethernet, or 10 GbE. 10 GbE will have the name IEEE 802.3ae with a timescale to deliver by March 2002. The philosophy and justification remain the same. If numerous users are generating data at 1 Gb/s or even many users at 100 Mb/s then even a 1 Gb/s backbone will soon become overloaded.

Several methods are up for discussion with the aim of getting at least 300 m transmission distance over multimode and tens of kilometres over singlemode. Unfortunately existing or legacy multimode fibre does not have the bandwidth to cope with a straightforward

10 Gb/s data stream sent down it. A figure of only 65 m transmission distance has been suggested if legacy fibre should be used this way. This would be for 50/125 working at 850 nm with a VCSEL laser; 100 m should be obtainable with a Fabry Pérot laser working at 1300 nm. The legacy fibre options are:

- Straightforward coding with 850 nm laser on legacy fibre, 65 m.
- Straightforward coding with 1300 nm laser on legacy fibre, 100 m.
- Five level optical coding on legacy fibre, 100 m.
- Parallel optics, i.e. 2.5 Gb/s sent down four separate fibres, 300 m.
- WDM, i.e. 2.5 Gb/s sent down the same fibre but four different wavelengths or 'colours', 300 m.

Another option is to introduce a brand new multimode fibre, laser launch optimised and with a much higher bandwidth. This would give a new 50/125 fibre a 300 m range with the low cost VCSEL.

The final option is of course to use singlemode fibre. There will be lower cost options for a few kilometres range and 1550 nm options for tens of kilometres range.

## Token Ring

Token Ring differs from Ethernet in that a 'token' is passed around a logical ring and only the station in possession of that token is allowed to transmit. This makes Token Ring *deterministic*; i.e. it is possible to predict network performance against load because one can easily calculate how long it will take the token to travel all the way around the ring. Ethernet is called *probabilistic* because one can only statistically assess the likelihood of stations wanting to transmit. The latter makes Ethernet performance very good under low loading conditions but poor when under high loading conditions and any application requiring real-time video or audio.

Token Ring was introduced in 1985 by IBM for use on 150 Ω STP cable and running at 4 Mb/s. This speed was later raised to 16 Mb/s and Token Ring was standardised in 1988 as IEEE 802.5.

Token Ring has lost much marketing ground to Ethernet and an attempt to regain some position was made with the introduction of some higher speed versions:

- 802.5t — 100 Mb/s operation over twisted pair copper cable.
- 802.5u — 100 Mb/s operation over two multimode optical fibres.
- 802.5v — 1000 Mb/s operation over two optical fibres.

## ATM

Asynchronous transfer mode (ATM), is a cell based switched technology. All ATM cells are 53 bytes long, with five bytes for addressing and the remaining 48 bytes for the payload.

ATM can be used as a to-the-desk LAN or as a wide area network or both. It can be used to link sites operating different LANs. It is however perceived as expensive and complex compared to Ethernet and as yet has only achieved a market penetration of less than 10% in the LAN environment.

Because of the simple asynchronous cell structure it is very easy to scale ATM in speed. Current versions include 25, 51 and 155 Mb/s over copper cable (both category 3 and category 5 versions are available) and 155, 622, 1200 and 2400 Mb/s over optical fibre. If category 6 cabling becomes more prevalent then the optical systems could probably be mapped onto a copper cable platform very quickly.

ATM does not carry a large overhead of error correcting codes, and presumes that the communications channel will mostly be error free. The ATM specification calls for better than 1 in $10^{10}$ bit error rate. The ATM copper cable variant also calls for an induced noise value of less than 20 mV picked up on the cable. A good quality cabling system is thus essential for this technology.

ATM is a connection-oriented network. This means that a virtual point-to-point link is set up between terminals before transmission can start. Ethernet and Token Ring effectively broadcast messages to everybody connected to the LAN and expect the correct station to pick out packets addressed to it. Quality of service (QoS) is seen as one of ATM's strongpoints and because of its asynchronous, cell-based, connection-oriented style it is considered as the best performer for delay-sensitive applications such as real time video, video conferencing and audio.

# FDDI

Fibre distributed data interface (FDDI), was developed as a high-speed optical backbone network for linking up departmental LANs running 10 Mb/s Ethernet and/or 4 or 16 Mb/s Token Ring. It was accepted as an ANSI (American National Standards Institute) standard in the mid 1980s.

At that time 100 Mb/s was perceived as more than adequate to link departmental LANs which in themselves could only develop an aggregate traffic load of 10 or 16 Mb/s maximum; and with the computers available in the 1980s even this wasn't thought likely to happen too often.

FDDI was designed to work on a token-based, dual optical ring; i.e. it requires four (62.5/125) optical fibres. This makes FDDI very secure and robust. If the ring is broken at any point then having four fibres allows an effective logical ring to be immediately formed on the remaining fibres. The token ring concept used is similar to the Token Ring LAN except that FDDI allows multiple tokens to circulate simultaneously, thus speeding up operation.

The maximum number of dual-attached (i.e. four fibre) devices connected to the FDDI ring is 500 and the total span can be up to 100 km. The maximum distance between devices is 2 km.

A single mode fibre version has been introduced called SMF-PMD which increases the distance allowed between devices from two kilometres to sixty kilometres. A category 5 copper cable variant, working up to 100 m has also been developed. It is called TP-PMD or twisted pair, physical medium dependent. This variant is also sometimes referred to as CDDI, or copper distributed data interface.

If the copper version of FDDI had been produced sooner and if one and ten gigabit versions of the backbone optical ring had been forthcoming then FDDI would still be seen as a viable corporate backbone LAN. It has unfortunately totally lost ground to Ethernet, which is considered to offer such an easy and logical upgrade path through 10, 100, 1000, and 10000 Mb/s. There is still a lot of FDDI equipment around however with 1999 surveys pointing to

20% of corporate users still operating FDDI. The FDDI specification requirement for 62.5/125 multimode fibre has also been seen as the benchmark performance for LAN fibre for most of the last decade.

# 6.5   Optical networks and channels

In the communications engineering sense, a 'channel' means the communications path between two communicating devices, and we would think of cables or radio to achieve the link. In the world of computers there is a subtle difference between 'networks' and 'channels'. LANs were developed to communicate between general purpose servers and workstations, i.e. the client-server model. Although LANs are very good at connecting large amounts of users they are not optimised for transferring large quantities of data, especially between mainframes. Links between mainframes and their high speed peripherals are called 'channels' in this context. We thus have a family of channel protocols which do a different job from LANs.

Channel protocols, such as SCSI and Bus & Tag, started off as

| Table 6.5 Optical LANs and channels | |
| --- | --- |
| Channel | Speed (Mb/s) |
| ESCON | 136 |
| Serial HIPPI | 800 |
| HIPPI-6400 (GSN) | 6400 |
| Fibre channel | 133–1062 |
| Optical LANs | Speed (Mb/s) |
| Ethernet 10baseF | 10 |
| Ethernet 100baseFX | 100 |
| Ethernet 1000baseSX/LX | 1000 |
| Ethernet 10GbE | 10000 |
| FDDI | 100 |
| ATM | 155–2400 |
| Token Ring IEEE 802.5v | 1000 |

short distance links but have evolved into longer distance, very high speed channels such as ESCON and fibre channel. The short distance channels such as SCSI and the old IBM Bus & Tag require dedicated copper cables but the optical links will be expected to run over the same optical backbone cabling as any other optical LAN. The cable network designer must therefore take into account the cabling requirements of LAN backbones such as FDDI, 100baseFX. 1000baseLX and 1000baseSX along with the channel requirements of systems such as ESCON, HIPPI and fibre channel. Table 6.5 summarises the current offerings.

## Small computer system interface (SCSI)

SCSI began as a means to connect peripherals to PCs and workstations at speeds up to 5 Mb/s. SCSI interfaces require their own copper cables but Table 6.6 demonstrates how SCSI has evolved to offer greater speed, more connected devices and other features.

For SCSI 3, the SCSI 2 standard was divided into a family of standards that will allow different physical transport layers to be defined; so it could be transported over fibre channel, for example.

Table 6.6  SCSI architecture

| SCSI architecture | Bus speed (Mb/s) | Max no. of devices |
|---|---|---|
| SCSI 1 | 5 | 8 |
| Fast SCSI | 10 | 8 |
| Fast/wide SCSI | 20 | 16 |
| Ultra SCSI | 20 | 8 |
| Wide ultra SCSI | 40 | 16 |
| Ultra2 SCSI | 40 | 8 |
| Wide ultra2 SCSI | 80 | 16 |
| Ultra3 SCSI | 160 | 16 |

## High performance parallel interface (HIPPI)

Originally there was 'parallel' HIPPI but now we have 'serial' HIPPI for longer distances. HIPPI was developed in the 1980s as a mainframe to mainframe I/O connection. It is now an ANSI standard, X3T9.3/90-043. HIPPI offers a point to point link at 800 or 1600 Mb/s over a 50-pair copper cable for a maximum length of 25 m. Serial HIPPI can send 800 Mb/s over 1 km of multimode or 10 km of singlemode optical fibre. A new project, started in 1997, is HIPPI-6400, also known as GSN, or gigabyte system network, which works at 6.4 Gb/s.

## Fibre channel

Fibre channel is standardised as ANSI X.3230-1994. It offers speeds from 133 to 1062 Mb/s with expansion planned to 2.12 and 4.24 Gb/s. Up to sixteen million nodes can be addressed with link lengths of up to 10 km. Fibre channel does not have a regular topology such as in Token Ring or FDDI but uses instead a 'fabric' which all users can attach to. Fibre channel can thus work point to point between two devices, with a central cross-point switch between many users or as an 'arbitrated loop' between lower-speed users. Fibre channel likes to present itself as a generic transport mechanism with a multi-functional set of layers. The highest layer, FC-4, allows other channels and networks, such as IPI, SCSI, HIPPI, IP and ATM to communicate via fibre channel.

## Enterprise system connection (ESCON)

ESCON is a trademark of IBM. IBM introduced ESCON in the early 1990s as a replacement to the old Bus & Tag system of connecting mainframes and peripherals. It offers high-speed links, 17 Mbytes/s, over fibre optic cable. Processors and peripherals communicate with each other via an ESCON director. A further development is FICON. This technology uses fibre channel at the lower level and ESCON is treated as the protocol above it. Physically, ESCON uses its own IBM proprietary duplex optical connector and a choice of 50/125, 62.5/125 and singlemode fibre. Cables are either 'trunk' cables, whereby there is only a fibre specification, and 'jumper' cables, which come in 4 metre to 500 metre lengths. The ESCON physical plant specification is in Table 6.7. See *Planning for Enterprise System*

Table 6.7 ESCON optical cabling design rules

| Optical Fibre or component | Attenuation at 1300nm | Bandwidth at 1300nm | Total link loss (dB) |
|---|---|---|---|
| 50/125 fibre max range-2km | 0.9dB/km | 800MHz.km | 8.0 |
| 62.5/125 fibre max range-3km | 1.0dB/km | 500MHz.km < 2km 800MHz.km for 2–3km | 8.0 |
| Singlemode fibre max range-20km | 0.5dB/km | N/A | 14.0 |
| Connector, multimode | 0.7dB | N/A | N/A |
| Connector, singlemode | 0.35dB | N/A | N/A |
| Splice, fusion | 0.4dB | N/A | N/A |
| Splice, mechanical | 0.15dB | N/A | N/A |

*Connection Links*, IBM publication GA23-0367-02, 3rd edition, for full details.

## Fibre channel and storage area networks (SANs)

Fibre channel is a high-speed network designed solely for fast communications between mainframes and bandwidth intensive peripherals such as large disk drives. Much of the physical interface technology of gigabit Ethernet came from existing fibre channel standards. There are a few more of these types of network around such as IBM's proprietary ESCON system and HPPI (high performance parallel interface).

A recent concept in the world of computer networking is the storage area network, or SAN. The main difference between a SAN and a more conventional LAN is that in a SAN, memory is no longer attached locally to a server. Different servers can now share the same storage and backup devices over long distances via an independent high-speed dedicated network. Twenty-five per cent of the world multi-user storage market is expected to be SAN based by 2002 and fibre channel appears to be the network

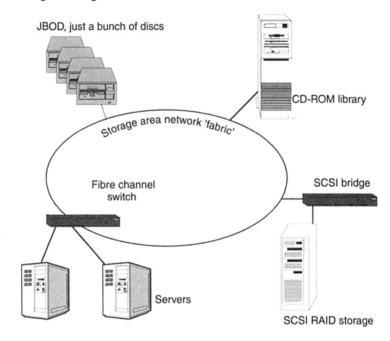

**Fig. 6.7** Storage network architecture.

technology of first choice to implement this architecture. Figure 6.7 shows an example of the SAN topology. To help overcome hardware compatibility problems the Fibre Channel Industry Alliance (formerly the Fibre Channel Association) has been formed, but for overall SAN management standards the Fibre Alliance was formed in February 1999.

## Optical performance requirements of optical LANs

LANs based on optical cable rather than copper cable offer greater bandwidth and distance capabilities but require more in-depth design knowledge on the part of the network designer. Optical LANs can be implemented as to-the-desk, backbone or campus networks or can be specialised mainframe-to-mainframe links only, such as IBM's proprietary ESCON (enterprise system connectivity), or fibre channel, the high speed link upon which gigabit Ethernet is based. Distance capabilities can range from a few hundreds of metres to tens of kilometres and so their capabilities must be understood.

Most LANs, including optical varieties, are attenuation limited, i.e. the signal is absorbed by the cable until it reaches a point where the receiver cannot pick it out from the background noise. Others are bandwidth limited. The classical example here is optical gigabit Ethernet, 1000baseSX and 1000baseLX. These protocols are limited by the available bandwidth of the multimode optical fibre used, e.g. a distance of only 220 m is foreseen when transmitting 1000baseSX at 850 nm over 62.5/125 standard multimode fibre.

Because the structured cabling standards allow multimode links of up to 2 km or 3 km for singlemode, there is often a mistaken assumption that all LANs will therefore work over that distance. Table 6.8 gives the attenuation budget and distances allowed per application according to the kind of fibre used.

Table 6.8 Optical budget and distances allowed for optical LANs

| Application | Wavelength (nm) | Loss budget (dB) | Distance 50/125 (m) | Distance 62.5/125 (m) | Distance single-mode (m) | Distance plastic fibre (m) |
|---|---|---|---|---|---|---|
| 10baseFL | 850 | 10 | 1340 | 2000 | | |
| 100 baseFX | 1300 | 11 | 2000 | 2000 | | |
| 1000baseSX | 850 | $2.38^{62.5/125}$ $3.37^{50/125}$ | 500 | 220 | | |
| 1000baseLX | 1300 | 2.35 | 550 | 550 | 5000 | |
| ATM 155 | 1300 | 7 | 2000 | 2000 | | 50 |
| ATM 622 | 1300 | 6 | 500 | 500 | 15000 | |
| ATM 155 swl | 850 | 7.2 | 1000 | 1000 | | |
| FDDI | 1300 | 11 | 2000 | 2000 | | |
| Token Ring | 850 | 13 | 1400 | 2000 | | |
| Fiber channel 133 | 850 | 6 | 2000 | 2000 | | |
| Fiber channel 266 | 850 | 6 | 2000 | 700 | | |
| Fiber channel 531 | 850 | 6 | 1000 | 350 | | |
| Fiber channel 1062 | 850 | 6 | 500 | 300 | | |
| Fiber channel 1062 | 1300 | 9.5 | | | 10000 | |

## 6.6   Video applications over structured cabling

LANs are not the only application for structured cabling systems. Other requirements are voice/telephony, building control/automation systems and video. Telephone systems are not particularly demanding and require only 3 kHz of bandwidth. Most PABX systems will happily transmit over several kilometres of category 5 cabling. Building control systems, and all those other applications grouped under the heading 'intelligent building systems' have similarly trivial channel requirements. Building control systems generally have very low data rates and more often nowadays they are made to work on Ethernet 10baseT interfaces anyway. Video however can be quite different. Any application involving pictures, moving or otherwise, will be a notoriously large consumer of bandwidth. There are different types of video application that may be called to run over structured cabling.

### Simple analogue baseband video

Analogue baseband video requires a bandwidth of 6–8 MHz. A typical source of such a signal could be a security CCTV camera. These cameras are usually made with a 75 Ω bnc coaxial connector output. A 75–100 Ω video balun is therefore required to connect this signal to the structured cabling and another balun will be required to convert the signal back again at the monitor end. Category 5 cable will typically support this signal over 300 m.

## Broadband analogue video

One or more television signals may be modulated onto a VHF carrier that will spread their spectrum up to 550 MHz. CATV is typical of this and coaxial cable was the original native media. Even with a video balun, category 5 cabling will struggle with this kind of frequency and transmission distances of around 65 m have been reported. Category 6 cabling will cope much better with broadband video and category 7 will be the best.

## Digital video

Directly digitising broadcast television signal through an analogue-to-digital converter is very bandwidth exorbitant. With a bandwidth of 8 MHz, the Nyquist sampling rate will be 16 MHz, and if encoded in 8-bit words the minimum line rate will be $16 \times 8 = 128$ Mb/s. With all the overheads associated with digitisation the actual rate for American standard NTSC colour pictures is 166 Mb/s and for European PAL it is 199 Mb/s. High definition television (HDTV), requires 1.5 Gb/s.

Much of the data in a television picture is redundant however. Video can be compressed by algorithms that look for changes from one picture to the next and only transmit the changes. The Motion Pictures Expert Group has developed a standard, MPEG-2, which compresses digital video[1]:

- High definition HDTV, 19.4 Mb/s.
- Broadcast, DVD quality, 10.8 Mb/s.
- Standard broadcast quality, 7.2 Mb/s.
- Near broadcast quality, 5.4 Mb/s.
- Medium quality videoconferencing, 4.3 Mb/s.
- Limited videoconferencing, 2.7 Mb/s.

Digitising a video stream is one thing but unless it is being sent across a dedicated point-to-point link then at some stage the data must be incorporated into, and transported by, a LAN. It is obvious that data streams of 5–19 Mb/s will soon eat up 10 and 100 Mb/s Ethernet channels and the concept of quality of service (QoS), comes into play. Video and attached coordinated audio are sensitive to network delay in a way that ordinary data packets are not. In an Ethernet system, packets will just queue up and wait if the network becomes congested and the end-user will be unaware of this unless the system becomes seriously congested. If it was a video link however the user would see frames dropped out and picture freezing and the quality would be deemed unacceptable. Even the act of putting video onto 10 or 100 Mb/s Ethernet will cause network congestion for all users.

The amount of delay acceptable has been defined by the ITU (International Telecommunications Union) as a maximum of 150 ms each way. The delay problem will be compounded if two LANs are connected by a wide area network working at a much lower speed than the LAN, e.g. 2 Mb/s. The extra delay caused by packets queuing at either side of this narrow 'bridge' is called serialisation delay.

Some LANs have QoS built in. ATM is the best for this and offers service levels defined as CBR (constant bit rate) which offers constant guaranteed bit rate to sensitive applications such as video, then ABR (available bit rate), VBR (variable bit rate) and finally UBR (unspecified bit rate). ATM is assisted in offering better video quality because it establishes a virtual point-to-point link before transmission starts, unlike Ethernet which is essentially a broadcast system.

FDDI and Token Ring have no native QoS capabilities and will always struggle with video. Ethernet will be able to offer some QoS

Table 6.9 Maximum number of video sessions on selected topologies

| Video bandwidth (Mb/s) | Ethernet (10 Mb/s) | Fast Ethernet (100 Mb/s) | ATM (155 Mb/s) | ATM (622 Mb/s) | Gigabit Ethernet (1000 Mb/s) |
|---|---|---|---|---|---|
| 10.8 | 0 | 9 | 14 | 57 | 92 |
| 7.2 | 1 | 13 | 21 | 86 | 138 |
| 5.4 | 1 | 18 | 28 | 115 | 185 |
| 4.3 | 2 | 23 | 36 | 144 | 232 |
| 2.7 | 3 | 37 | 57 | 230 | 370 |

functions and it will be helped out by the availability of faster and faster connections, i.e. 10, 100, 1000 and 10000 Mb/s. Ethernet works on a broadcast system whereby every user connected can see the data packets. If video users are kept on the same segment then network congestion can be localised to some extent. Using internet protocol (with high speed Ethernet as the data link/physical layer) then a CoS or class of service can be achieved. This includes multi-cast video whereby only selected groups of users receive the video traffic rather than broadcast, where every user receives the traffic. Table 6.9 gives the maximum number of video sessions on selected topologies[1].

## RGB video systems

Although video can be digitised and transported on a LAN it is still seen as expensive and heavy on bandwidth. Many users still prefer to use a dedicated point-to-point or even switched analogue video system called RGB. RGB means splitting the colour signal into its three primary colours, i.e. red, green and blue. The receiver re-combines the three colour signals taking its synchronisation signal usually from the green channel. On a structured cabling system three of the four pairs carry the three colour signals. RGB on structured cabling has replaced some of the more established financial infor-mation distribution systems such as Reuters and Bloomberg, which used to require their own dedicated cabling.

A problem that RGB signals have is differential delay or asymmetric skew. Each colour component must arrive at the receiver at the same time if the original signal is to be reproduced. If they arrive separated out in time then the resulting picture will have annoying colour fringes. The latest cabling standards, such as category 5e, category 6 and category 7 all have differential delay requirements of 50 ns or less. This figure basically comes from the requirements of gigabit Ethernet, but RGB video needs 20 ns differential delay or better. It is possible to buy delay lines, which will slow down the fastest signal to the rate of the slowest signal, but this is an expensive option that needs setting up and subsequent recalibrating. It is far better to specify a cable with a differential delay of 20 ns or less over a 100 m cabling link.

# Reference

1   Fritz J, 'Caught up on video', *Data Communications*, October 1999 (quoting in turn a University of West Virginia study of February 1999).

# 7

# Copper cable technology — cable

## 7.1 Introduction

There are several different types of copper cable constructions. As with most choices in life there are pros and cons for each type. Modern structured cabling for voice and LANs is based on twisted pair cables whereas most video signals still travel on coaxial cable, or coax as it is usually abbreviated. All the common types of copper communications cable will be described here. Other conductor materials are sometimes used in the cable industry, such as copper clad steel, aluminium and high strength alloys, but these are usually for specialist applications such as military or aerospace installations. This book deals only with cables made from annealed copper, sometimes referred to as PACW (plain annealed copper wire) or TACW (tinned annealed copper wire). The four major groups of copper cables are:

- Multicore.
- Twisted pairs.
- Quads.
- Coax.

## 7.2 American wire gauge

An important parameter to know is the diameter of the copper wire in use. This may be expressed in millimetres, e.g. 0.5 mm or more

| Table 7.1 Gauge size of copper wire | | | |
|---|---|---|---|
| Gauge size | Diameter (mm) | Weight (kg/km) | dc resistance ($\Omega$/km) |
| 29 | 0.3 | 0.64 | 240 |
| 28 | 0.32 | 0.7 | 210 |
| 26 | 0.4 | 1.13 | 135 |
| 24 | 0.5 | 1.84 | 85 |
| 22 | 0.63 | 2.82 | 85 |
| 19 | 0.9 | 5.65 | 54 |
| 17 | 1.2 | 10.2 | 16 |
| 16 | 1.3 | 11.3 | 14 |

commonly in the datacomms industry by the acronym AWG. AWG stands for American wire gauge. Contrary to popular belief there is no international agreed standard for the hard metric equivalent for AWG. American wire gauge started off as a means of stating how many times a 0000 gauge of 0.46-inch diameter copper bar was drawn down. Thus 24 gauge means that the bar has been drawn down 24 times. Gauge sizes have become an industry convention so that 24 AWG is generally 0.5 mm (0.0253 inches) in diameter. But there is no guarantee that that 24 gauge from one manufacturer will be exactly the same as from another manufacturer. Table 7.1 gives the most commonly used sizes of copper wire.

# 7.3  Solid and stranded conductors

Solid conductors are commonly used in cables such as twisted pair data cable as they are less costly to make and have a smaller cross sectional area than stranded conductors. Solid conductors however are not good at repeated bending such as might be experienced by a patchlead. So main horizontal and riser cabling, which will probably only ever be flexed when it is installed, is made of solid conductors.

Stranded conductors are typically made of seven smaller wires (typically expressed as 7/0.203, which is equivalent to 24 AWG) whose total cross sectional area may add up to the same as a solid conductor of the same gauge size. The real cross sectional area will be larger because of the interstitial gaps between the smaller conductors contributing to the overall cable size. Also very large cables, such as welding cables are stranded because otherwise they would be impossible to bend at all. For structured cabling the typical wire sizes for patchlead are 24 and 26 AWG.

Note that there is sometimes a difference of terminology with stranded cables:

- Patchcord and patchcable usually mean unterminated lengths of flexible stranded-conductor cable.
- Patchlead usually means a short length of flexible cable terminated on both ends with a connector.

Warning! Manufacturers and customers alike regularly mix up these terms.

# 7.4  Insulation material

The copper wire must have a layer of insulation over it. The electrical parameters of the insulation material play a critical part in the overall performance of the cable. Power or energy cables are typically insulated with PVC, but this has too poor a performance for high speed LAN cable. Note again the terminology. The insulation goes around the copper conductor. It has a crucial role in the cable's electrical performance. A sheath or jacket goes around the whole cable for mechanical/physical protection. It plays a very minor part in the electrical performance of the cable. So a high performance LAN cable may have one material for the insulation and still use PVC for the sheath. The insulation material is sometimes referred to as the dielectric. The dielectric has four properties of interest:

- Dissipation factor.
- Dielectric strength.
- Insulation resistance.
- Dielectric constant (permittivity).

The dissipation factor is the ratio of the conductance of a capacitor to its susceptance. It gives a measure of power loss through the dielectric and is worse at higher frequencies.

The dielectric strength is a measure of how high a voltage the dielectric can withstand before it breaks down. It is defined as the voltage at breakdown divided by the thickness of the insulating material at the point of breakdown, i.e. Volts/mm. The insulation resistance is length dependent, usually expressed as MΩ/km.

The dielectric constant is a measure of the relative ability of a material to hold a charge, like a capacitor, compared to air. A low figure, as close to one as possible, is desirable, because capacitance in a cable is generally an unwanted quantity. Capacitance takes energy and time to charge up and limits high frequency performance.

From the dielectric constant we derive the NVP or nominal velocity of propagation. This is the ratio of the speed of the signal in a cable relative to the speed of light in a vacuum:

$$NVP = \frac{1}{\sqrt{\text{dielectric constant}}} \qquad\qquad [7.1]$$

Properties of common cable making materials are shown in Table 7.2.

Table 7.2 Properties of common cable making materials

| Material name | Abbreviation | Dielectric constant |
| --- | --- | --- |
| Cellular PTFE | CPTFE | 1.3 |
| Cellular polypropylene | CPP | 1.3 |
| Cellular low density polyethylene | CLDPE | 1.4 |
| Polytetrafluoroethylene* | PTFE | 2.0 |
| Fluoroethylene polymer* | FEP | 2.1 |
| Low density polyethylene | LDPE | 2.2 |
| High density polyethylene | HDPE | 2.3 |
| Polyvinyl chloride | PVC | 6.0 |
| Polyurethane | PUR | 7.5 |

*PTFE and FEP are often referred to by their DuPont trade name of Teflon.

Apart from the dielectric constant we have the variables of cost, ease of processing, dissipation factor, operating temperature, flammability, insulation resistance, resistance to water and oil based solvents and resistance to ozone and UV light.

All these variables have to be taken into account when choosing the right insulation material. We can see that the electrical performance of PVC and polyurethane make them unsuitable as insulators but their cost and mechanical performance make them ideal as sheath materials. PVC is the most commonly used sheath material for indoor cabling, despite concerns over its flammability. Its low resistance to water and UV light make it generally unsuitable for outdoor use. The most common sheath material for outdoor applications is carbon-loaded polyethylene.

Air has the best dielectric constant (after a pure vacuum) and we can see from Table 7.2 that injecting small bubbles of gas or air into the insulation to give a foamed or cellular effect greatly improves the dielectric constant of the material. The gas is mechanically injected into the liquefied insulant when it is being extruded over the conductor, or a chemical that decomposes into a gas can be mixed in.

The most common insulating material for LAN cables is polyethylene or polypropylene, either foamed or solid. These two polymers are often broadly referred to as polyolefins.

# 7.5   Fire rating of cables

Although PVC is the most common sheathing for indoor data cables, there is a choice of fire retardant PVC and a range of low flammability, zero halogen materials now available. Such materials have many interchangeable acronyms, such as LSF (low smoke and fume), LSFOH (low smoke and fume, zero halogen) and LSOH (low smoke, zero halogen).

The halogens are a class of elements consisting of: chlorine, fluorine, bromine, iodine, astatine. They are added to many plastics to act as stabilisers and flame-retardants. Their combustion products give rise to many halogenated, acidic compounds such as hydrochloric acid. The acidic fumes have a very toxic effect on people and a very damaging effect on electronic equipment such as printed circuit boards. Many users, especially in Europe, are trying to do away with halogenated cables within buildings. Non-halogenated materials will still burn and give off noxious fumes such as carbon monoxide, but no halogenated, acidic gasses. A number of IEC standards relating to fire performance issues are:

- IEC 332-1 — flammability test on a single burning wire.
- IEC 332-3-c — flammability test on a bunch of wires.
- IEC 754 — halogen and acidic gas evolution from burning cables.
- IEC 1034 — smoke density of burning cables.

In America there are some very strict flammability standards for indoor cabling. The highest test is for cables placed in the plenum area. This is any area within a building that carries environmental air, such as the return airflow from an air-conditioned area. The moving air would make any fire in these spaces very dangerous and easily spread. A plenum space might typically be between the ceiling tiles and the deck of the floor above. Down from the plenum rated cables come riser grade cables, and below that general purpose cables. There is a substitution rule that allows plenum rated cables to be used in the riser, but riser cables could not be used in the plenum zone. Plenum cables have two disadvantages:

1  They are expensive, typically three to four times the cost of PVC jacketed cables.
2  They are expensive because they are jacketed with PTFE/FEP (Teflon) types of materials. These do of course contain a halogen, fluorine, so when they do eventually burn they will give off halogenated, acidic gasses.

The American National Electrical Code (NEC) requires cables to be marked according to their classification, as shown in Tables 7.3 and 7.4, and permits substitutions of higher rated cables into lower rated environments (NEC tables 770-50, 770-53, 800-50, 800-53). Some relevant American fire tests are:

Table 7.3 NEC cable marking

| Cable marking | Type |
|---|---|
| MPP | Multipurpose plenum |
| CMP | Communications plenum |
| MPR | Multipurpose riser |
| CMR | Communications riser |
| MP,MPG | Multipurpose |
| CM,CMG | Communications |
| CMUC | Undercarpet |
| CMX | Communications, limited |

Table 7.4 Permitted cable substitutions

| Cable type | Permitted substitution |
|---|---|
| MPP | None |
| CMP | MPP |
| MPR | MPP |
| CMR | MPP,CMP,MPR |
| MP,MPG | MPP,MPR |
| CM,CMG | MPP,CMP,MPR,CMR,MP,MPG |
| CMX | MPP,CMP,MPR,CMR,MP,MPG,CM,CMG |

- UL 910    plenum test.
- UL 1666   riser test.
- UL 1581   general purpose.

## 7.6   Screening or shielding of cables

The terms screening and shielding are synonymous and inter-changeable. American manufacturers tend to use the word shielding and British manufacturers the word screening. This book will adopt the convention — screening. Both words refer to the practice of enclosing the conductors in a cable within a conducting and earthed (grounded) element for the purpose of preventing electromagnetic fields from entering or leaving the cable. The relevance and efficacy of screening data cables is a controversial subject. Some countries such as Germany install over 90% screened cable in their market. In America over 95% of the market for premises cable is for unscreened. In Britain only 10% of the market is for screened cables.

The electromagnetic effects of screening are discussed in other parts of this book. In this section we will concern ourselves with the mechanical methods of screening.

In general, the more metalwork put around a cable then the better the screening effect will be. A copper tube around a cable will perform better than a thin aluminium foil tape but more screening elements bring many disadvantages:

- More screening means a more expensive, heavier and larger cable.
- More screening needs more installation effort to terminate.
- All screened cable need an effective earthing strategy.

The following screening methods are available in LAN/premises cabling and are shown in Figs. 7.1–7.4. *Note*: the terminology is not consistent between all manufacturers, installers and users. Check that all parties are talking about the same thing when discussing terms like STP, FTP etc to avoid major problems later on.

## UTP

Unscreened (unshielded) twisted pair (the 'base' model), Fig. 7.1.

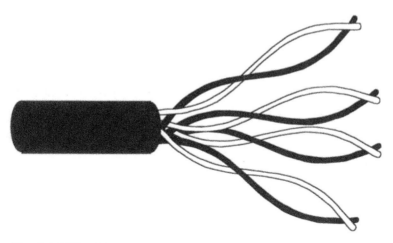

**Fig. 7.1** UTP cable.

## FTP

Foil screened twisted pair, Fig. 7.2. Aluminium foil is wrapped longi-
tudinally or helically around the copper pairs. A drain wire (an unin-
sulated wire running the length of the cable and in contact with the
conducting side of the foil to ensure a low impedance connection to
the foil throughout) is put under the foil or sometimes spirally wrapped
around the outside.

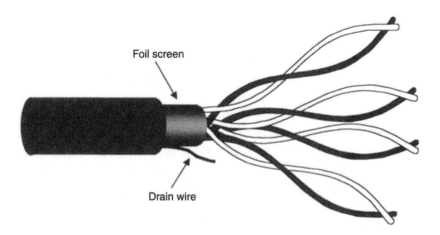

Foil screen

Drain wire

**Fig. 7.2** FTP cable.

## S-FTP

Screened-foil twisted pair, Fig. 7.3. An aluminium foil is wrapped around the copper pairs and then a tinned copper braid is applied around the foil. No drain wire is necessary as the foil and braid are in electrical contact.

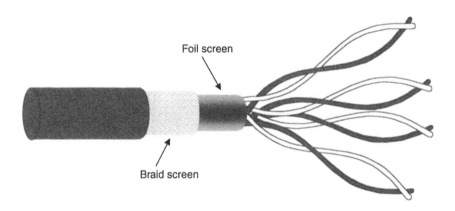

**Fig. 7.3** S-FTP cable.

## STP

Screened twisted pair, Fig 7.4. This usually means that each pair has an individual foil screen around it and the whole cable has a braid or foil and braid screen around it. Examples of such cables are the IBM type 1 cable and the category 7 cables. Another term sometimes used is PIMF (pairs in metal foil).

Another term sometimes used is ScTP, which has been used to describe any of the above.

For telecommunications there is a kind of cable called transverse screened which is a multipair designed for 2 Mb/s transmission. Transmit and receive pairs are separated by an 'S' shaped aluminium foil so that all transmit pairs are enclosed within one lobe of the 'S' and all the receive pairs reside in the other lobe of the 'S'.

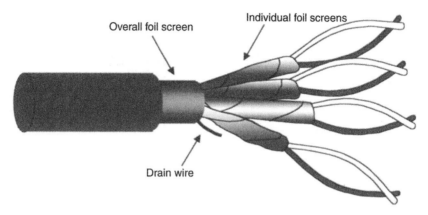

**Fig. 7.4** STP/PIMF cable.

# 7.7  Multicore cables

Multicore cables have each individual insulated conductor laid straight within the cable, i.e. there is no attempt to twist any of the conductors together. This makes for a low cost cable but it is only suitable for very low speed communications or low voltage power. The lack of twist between conductors would give unacceptable levels of crosstalk for any method of communication requiring a few hundred kilohertz of bandwidth or more. As a result of this drawback, multicores are limited to low speed applications such as telemetry, process control and other low speed requirements such as local printer connections. The standard known as RS 232 or V24 defines a low speed link, up to 19.6 kb/s, over a 25-wire multicore, where each conductor has an agreed signalling function. Note this is an *unbalanced* application as each signal conductor shares a common earth return conductor.

We can see from Fig. 7.5 that the transmission speed versus distance is very limited by today's standards.

Multicore cables can be screened or unscreened with usually between four and twenty-five conductors. But their low speed and distance limitations give them no place in modern structured cabling systems and Local Area Networks apart from local printer connection.

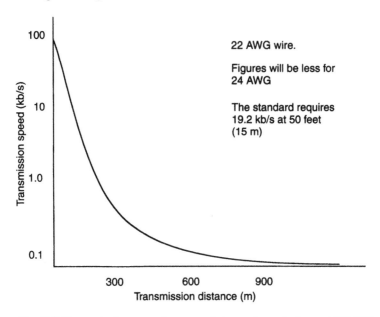

**Fig. 7.5** Transmission speed versus distance for unbalanced RS 232 circuits.

# 7.8   Twisted pair cables

A twisted pair cable gives much better transmission performance. To start with it is a balanced cable. This means that an equal and opposite current flows down each of the two conductors of the twisted pair, there is no common earth return for all the signals as in RS 232 multicore. As current flows down any conductor it creates an electromagnetic field around it. This field can be a cause of pickup, or crosstalk, into adjacent conductors, and can be a major source of interference. The higher the frequency of operation, the worse the crosstalk phenomenon becomes. By having an equal but opposite current flowing in each conductor, a large part of the external magnetic field is cancelled out, greatly improving the crosstalk performance of the cable (Fig. 7.6).

The twisted pair cable is also more immune to external noise and crosstalk from other conductors. Induced noise tends to be common mode, i.e. the noise current flows down each conductor, in the same direction and of the same magnitude. Several tricks in the receiver

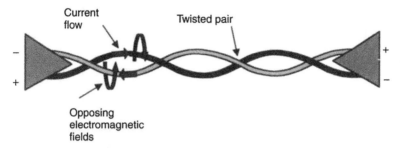

**Fig. 7.6** Current flow in balanced cable.

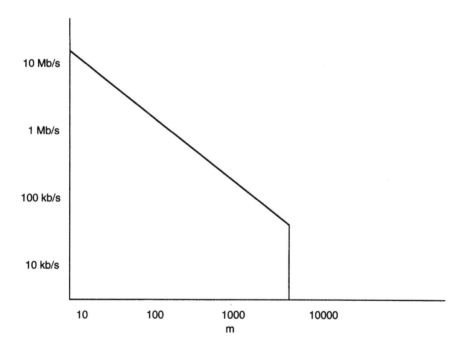

**Fig. 7.7** Improved transmission speed versus distance when using RS422 protocol with balanced cable.

design can cancel out this common mode noise as the receiver is only looking for differential signals on the two conductors of the pair, i.e. equal and opposite (Fig. 7.7).

Twisted pair is the dominant cable type in structured cabling and

Local Area Networks. There can be any number of pairs within a cable. Structured cabling generally uses four but sometimes twenty-five pairs of 24 AWG are used for backbone or zone distribution systems. Cables used for voice telephony frequently go up to very high pair counts, such as 4000 pairs, but often with smaller conductors.

As we have already mentioned, twisted pair cables can be screened (shielded) or unscreened (unshielded). Unscreened cable is usually referred to as UTP. The cable can be screened with an overall screen of aluminium foil or copper braid or even both placed around the pairs and under the sheath (jacket), when it is often referred to as FTP or S-FTP or ScTP. Each pair can also have its own screen of foil in which case the cable is often referred to as STP or PIMF. It should be noted however that users and suppliers alike often mix up this terminology.

Other manufacturing methods to improve cable performance are:

## Lay length

The amount of twist put in the pair. Usually the more twists applied to the pair then the better it will be for crosstalk performance; around 14–23 mm is common. Plus all four pairs are twisted together approximately every 100 mm. The limiting factors are (a) cost — more twist means more material and slower processing, and (b), subsequent rise in attenuation and signal delay.

## Staggered lay

A different amount of twist, or lay length, is put into each pair within the cable, or else pickup could occur between the pairs over very long cable runs.

## Differing pair sizes

Some manufacturers put in slightly higher diameter wires on the two worst performing pairs. This raises the whole cable performance without too much of a cost penalty.

## Foamed dielectric

As air is a very good dielectric material, the cable performance can be improved by turning the normally solid insulation material into a foam structure. This is achieved by gas injection at the insulation extrusion stage or by adding a chemical that gives off a gas into the liquid insulation material.

## Cross-linked polymerisation, XLPE

The insulation material performance can be improved by cross-linking the molecules within the insulation to form larger molecules. This is normally used for very high performance cables where cross-linking allows a high cable performance with thinner sheath materials and is used in applications where cable physical space is at a premium, e.g. telephone exchanges and aeroplanes. The cross-linking can be achieved chemically or by bombarding the cable with a high intensity electron beam.

## Differing insulation materials

Some manufacturers serving the American market have mixed in different insulation materials to obtain lower cost cables that still meet the UL plenum test, for example two pairs of PTFE and two of polyolefin. This practice is not acceptable nowadays as the differing pairs can have radically different NVPs (nominal velocity of propagation) which plays havoc with modern protocols that share the information between all four pairs, such as 100baseT4 and 1000baseT.

## Bonded pairs

The four pair cable is designed under the assumption that the pairs will always keep the same relative position to each other. After the rigours of installation this is not always the case. Some manufacturers bond the pairs together to keep that same relative position and maintain consistency of performance. It is also possible to obtain a flat four pair cable with the same design objectives in mind.

## Striated conductors

At high frequencies a phenomenon known as the skin effect comes into effect. At high frequencies the signal energy tends to travel near to the surface or skin of the conductor. The centre of the conductor plays little part in the transmission of energy. With less effective cross-sectional area available the attenuation goes up. The answer is to have a larger conductor or to increase the surface area of the conductor. The latter can be achieved by striating the conductor by pulling it through a die which puts tiny grooves into the surface, almost like a cog wheel effect but where the individual teeth are too small to be seen with the naked eye. This method gives a marked improvement in the stability of the cable impedance over the frequency range in use.

## Physical separation of pairs

There is no better way of improving crosstalk than by simply putting in some physical distance between the pairs. Most category 6 cables have adopted this approach by putting in a small plastic cross or former to hold the four pairs apart (see Fig. 7.8).

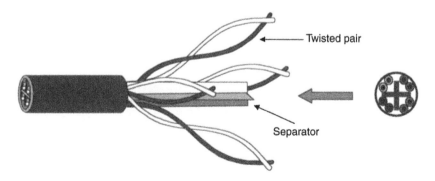

Twisted pair

Separator

**Fig. 7.8** Typical category 6 cable construction with physically separated pairs.

# 7.9    Categories and classes of cable

Since the mid 1990s, the cables used in structured cabling systems have been defined as categories and classes amongst the following standards, TIA/EIA 568A, ISO 11801 and EN 50173 (note, EN50173 does not recognise category 4). Table 7.5 gives the cable categories and corresponding frequency range.

The 'category' cables are all based on indoor grade, 100Ω, four-twisted pair cable although multipair such as 25-pair, composites, hybrids and bundled cables are also allowed. The standards recognise the following cables:

*TIA/EIA 568B — horizontal cabling*
4-pair 100Ω UTP/ScTP cable, Cat 3, 4 or 5 (5e recommended)
2-fibre 62.5/125 or 50/125 optical fibre cable.
2-pair 150Ω STP-A cable is recognised but not recommended for
  new installations.

*Backbone cabling*
100Ω twisted pair cable cat 3,4 or 5/5e
62.5/125 and 50/125 optical cable
single-mode optical cable.
75 and 50Ω coax are described in annexe C of 568B but are not
  recommended for new installations.

*ISO 11801 — horizontal cabling*
Preferred        100Ω balanced cable (screened or unscreened)
                 62.5/125 optical fibre
Alternative    120Ω balanced cable

| Table 7.5  Cable categories and frequency range | |
| --- | --- |
| Cable category | Frequency range (MHz) |
| Category 3 | 16 |
| Category 4 | 20 |
| Category 5 | 100 |
| Category 6 | 200 |
| Category 7 | 600 |

150 Ω balanced cable

50/125 optical fibre.

*Backbone cabling*

50/125, 62.5/125 or singlemode fibre

100, 120 or 150 Ω balanced cable.

European cables are also defined in the following publications:

EN 50167    Sectional specifications for horizontal floor wiring cables with a common overall screen for use in digital communications.

EN 50168    Sectional specifications for work area wiring cables with a common overall screen for use in digital communications.

EN 50169    Sectional specifications for backbone cables, riser and campus with a common overall screen for use in digital communications.

The above three standards are being replaced by the following:

EN 50288-2-1    100 MHz screened, horizontal and backbone.
EN 50288-2-2    100 MHz screened, patch.
EN 50288-3-1    100 MHz unscreened, horizontal and backbone.
EN 50288-3-2    100 MHz unscreened, patch.
EN 50288-4-1    600 MHz screened, horizontal and backbone.
EN 50288-4-2    600 MHz screened, patch.
EN 50288-5-1    200 MHz screened, horizontal and backbone.
EN 50288-5-2    200 MHz screened, patch.
EN 50288-6-1    200 MHz unscreened, horizontal and backbone.
EN 50288-6-2    200 MHz unscreened, patch.

Cables are also defined in the following American standards:

NEMA WC-63.1    Performance standards for twisted pair premises voice and data communications cable.

NEMA WC-63.2    Performance standards for coaxial communications cable.

NEMA WC-66    Performance standards for category 6, category 7 100 Ω shielded and unshielded twisted pair cables.

| ICEA S-80-576 | Communications wire and cable for wiring of premises. |
| ICEA S-89-648 | Aerial service wire. |
| ICEA S-90-661 | Individually unshielded twisted pair indoor cables. |
| ICEA S-100-685 | Station wire for indoor/outdoor use. |
| ICEA S-101-699 | Category 3 station wire and inside wiring cables up to 600 pairs. |
| ICEA S-102-700 | Category 5, 4-pair, indoor UTP wiring standard. |
| ICEA S-103-701 | AR&M riser cable. |

# 7.10   External cable

There is no provision in the major standards for external grade cables above category 3. Category 3 cable is still often used for external and internal telephone backbone wiring in very high pair counts. For high performance LAN links however it is recommended to use optical cable for interbuilding links. If it is required to use an external copper cable then it may be possible to run a standard category 5 within a conduit as long as it is totally protected from water and extremes of temperature. Internal grade cables (especially PVC sheathed) have a low tolerance to all fluids, UV light and temperature ranges lower than −10°C and higher than 60°C.

External grade cables are weatherproofed by one of the following measures:

• Carbon loaded polyethylene sheath. Polyethylene is naturally water-resistant and the carbon loading increases its tolerance to UV light.
• Aluminium foil moisture barrier under the sheath. When the foil is coated so that it is hermetically bonded then it provides a water-proof barrier.
• Petroleum jelly in the interstitial spaces of the cable. If water does penetrate the cable then the gel prevents it from travelling down the cable in the spaces between the pairs (the interstices). This cable is sometimes called filled PIC (plastic insulated cable).

- Air pressurisation. With no gel filling it is possible to pressurise the cable with dry air. If the cable is damaged then the positive air pressure will keep the water out. Also the drop in air pressure can be monitored at the central station and an alarm sounded. Cable buried in the ground may be pressurised up to 3 psi whilst cables within cable ducts may be pressurised up to 6 psi.

External cable routes may be direct buried, placed within underground cable ducts or aerial. Underground cables may also be armoured with steel wire or tape according to circumstances.

An aerial cable may be suspended between telegraph poles or directly between buildings. The cable may be self-supporting or it may be attached to a catenary wire (messenger wire) for strength. Aerial cables are subject to other problems such as severe temperature changes, ice loading, wind loading and even shotgun attack!

When metallic cables enter buildings any armour should be grounded at the entrance facility and over-voltage and over-current protectors inserted in series with the incoming circuit to protect people and equipment from surges caused by lightning or any other accidental high voltage contact. If using a high performance cable such as category 5 or above then it is imperative that the manufacturer of the surge arresting equipment can guarantee that the electrical performance of the cable will not be degraded by the insertion of such devices.

Outdoor cables are made to be weatherproof and as such they use materials such as polyethylene and petroleum jelly. Unfortunately these materials are far too flammable to be allowed within buildings. Most countries have fire codes limiting the distance into a building that an external cable is allowed to run before it is either spliced onto an internal grade cable or is enclosed within a metallic conduit; for example NEC 800-50 limits external grade cables to runs of 15 m or less when they enter buildings.

Users requiring external grade category 5 or higher cable should consult a number of manufacturers to see what products are currently available.

# 7.11   Quads

A quad is an alternative construction to a twisted pair cable. It is constructed of four conductors laid up and twisted together and can give a smaller cable diameter than the equivalent two pair construction. However an 8-conductor quad, equivalent to four twisted pairs would have to be a flat, figure-of-eight shape. The quads also have to be very accurately made to achieve the desired electrical performance and so can be expensive. For the above reasons quads are rarely seen in structured/LAN cabling.

# 7.12   Coaxial cable 'coax'

A coax cable, Fig. 7.9, has two conductors with one laid co-axially or concentrically to the other, hence the name. Coax is usually an unbalanced cable as the outer conductor has the effect of a screen and is earthed. The central conductor can be solid or stranded copper and the outer conductor can be aluminium or copper foil or copper braid or any combination. The two main parameters of coaxial cable are attenuation and characteristic impedance. Generally the larger the cable then the lower the attenuation will be. The performance of the cable may be improved by having several layers of braid

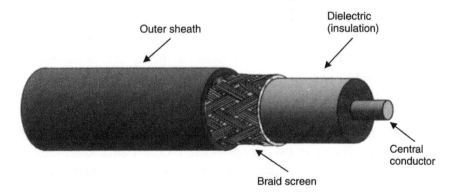

**Fig. 7.9** Coaxial cable.

screens or sandwiching foil in between braid screens. The insulation between the conductors is usually referred to as the dielectric and it may be solid or foamed or air spaced. The latter involves separating the two conductors with the minimum amount of plastic possible. This may involve a cellular construction or even a simple plastic 'wire' spiralling around the central conductor to separate the two conductors. Coaxial cables often have an industry standard generic title starting with RG or URM, e.g. RG62.

Coaxial cables have a very good high frequency, broadband performance and are still greatly used for video transmissions such as in CATV (community antenna television) distribution, CCTV/security (closed circuit television) and RGB (red, green, blue) graphic systems.

For distances of more than a few hundred metres optical fibre becomes more cost effective. CATV distribution is often planned as hybrid fibre/coax, i.e. the long runs from the head-end are on fibre but the signal is distributed to the individual users over the last 500 m or so on coax. For less than 100 m most users start to look at category 5 or higher, twisted pair. CATV requires over 500 MHz of bandwidth and category 5 cable struggles to accommodate this over 100 m. A security camera requires about 6 MHz bandwidth, and this signal is easily transported over 300 m on category 5 cable. Figure 7.10 shows the relative attenuation of coaxial cable compared to optical fibre, and it becomes obvious why optical fibre is so effective in distances beyond a few hundred metres.

The original Ethernet system was designed to run over coax and a later system using thinner, cheaper coax (Thinnet) was introduced soon after. All Ethernet systems today are designed to run over twisted pair or optical cable. Other computer systems that were designed to run over coax were the IBM AS400 and the IBM 3270. These computer systems are sometimes called legacy systems because they have been around a long time but they still do a good job. They can be connected to modern structured cabling systems by the use of a balun. A balun essentially converts one cable impedance to another, so an IBM 3270 front end processor, expecting to be connected to RG62 93 $\Omega$ coax cable can be connected to a 4-pair, 100 $\Omega$, category 5 twisted pair cable by using the appropriate

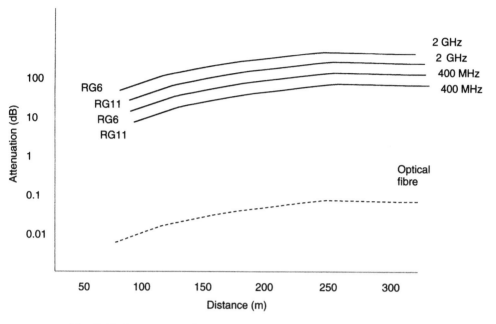

**Fig. 7.10** Attenuation of various grades of coax and optical fibre.

Table 7.6 Original coax cable media for some 'legacy' communications systems

| Application | Generic part number | Description |
|---|---|---|
| IBM 3270/5250/3790 | RG62 | 93 Ω impedance coax |
| IBM AS400 system 34/38 | | 105 Ω twinax |
| Ethernet | 'Thick coax' 10base5 | 50 Ω impedance coax |
| Ethernet | 'Thin coax' 10base2 | 50 Ω impedance coax |
| Racal planet | | 75 Ω twinax |
| Wang VS/OIS | | 75 Ω twinax |
| Reuters | RGB triax | Triple 75 Ω coax |
| Analogue video | RG6/11/59/216 | 75 Ω coax |

balun. The 3270 device is totally unaware that it is now communicating via twisted pair rather than coaxial cable. Table 7.6 shows some legacy coax requirements.

Another form of coax is called 'radiating coax' or 'leaky feeder'. In these products the outer braid is made with a very small coverage so that the high frequency signals contained within the cable can leak, or escape to a small extent. These cables are used for radio communications within tunnels as the coax leaks out small amounts of radio energy along its length, and conversely can absorb radio signals generated by users within the tunnel.

# 8

# Copper cable technology — components

## 8.1   Introduction

To be useable a cable has to be terminated in some form of connector. Various levels of interconnection may also be required in between the two end points of the cable system. There are many, many different kinds of connector. The choice of connector depends upon the electrical performance required, the environment in which it must be installed, number of conductors to be terminated, type of cable involved and power, voltage and current levels expected. A 600 V, twelve-core power cable connector for an exposed position on a North Sea oil rig will bear little resemblance to a connector for an indoor data cable containing four twisted pairs.

Coaxial cables require their own special connectors. A BNC style is commonly used in communications applications. Multipair cables often use 'D' type connectors, typically with 9 or 25 pins. The IBM cabling system, based on two-pair, screened 150 $\Omega$ cable uses its own special connector, known as the IBM or hermaphroditic connector because two can connect together; there is no male/female, plug or socket version. Structured cabling systems for Local Area Networks have now standardised on an 8 pin modular connector commonly known as an RJ45. RJ45 is a USOC (Universal Service

Order Code) designation for specific applications and/or manufacturers, and the correct generic title is IEC 60603-7, but RJ45 has come to be the commonly accepted term. This chapter will mainly concern itself with connecting hardware used in structured cabling systems.

### 8.1.1   Structured cabling system components

All the components, and the physical topology where they can appear, are defined within the standards. Figure 8.1 gives the ISO 11801 generic cabling system topology. Figure 8.2 gives the American TIA/EIA 568A/B equivalent.

The two generic cabling systems are almost identical in function but the terminology is slightly different (see Table 8.1). Within the premises, the cabling equipment termination and cross-connect equipment may be found in telecommunications rooms (or closets), equipment rooms or entrance facilities.

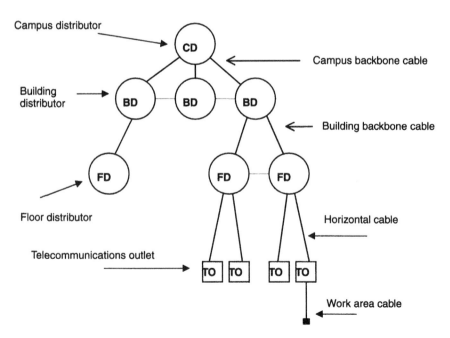

**Fig. 8.1** ISO 11801 generic cabling system.

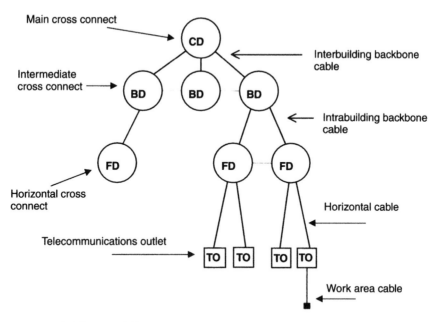

**Fig. 8.2** TIA/EIA 568A/B generic cabling system.

| Table 8.1 Generic cabling systems | |
|---|---|
| ISO 11801 functional elements | TIA/EIA 568 telecommunication cabling system structure |
| Work area cabling | Work area cabling |
| Horizontal cable | Horizontal cable |
| Floor distributor | Horizontal cross connect |
| Building backbone cable | Backbone cable |
| Building distributor | Intermediate cross connect |
| Campus backbone cable | Backbone cable |
| Campus distributor | Main cross connect |
| Telecommunications outlet | Telecommunications outlet |
| Transition point | Transition point |
| Consolidation point | Consolidation point |
| Multi-user telecom outlet assembly | Multi-user telecom outlet assembly |

The format of the hardware within the distributors and cross-connects comes in the shape of 19-inch patch panels or cross-connect strips. These in turn contain terminating hardware in the form of connectors such as RJ45 and IDCs.

# 8.2   Connectors

## 8.2.1   Insulation displacement connector

The conductors within a cable need a permanent, low loss connection to the terminals of the connector. This may be effected by soldering, crimping, screw terminal or spade and lug. Insulation displacement connectors (IDCs), were invented as a low-labour method of termination to replace earlier methods such as soldering and crimping.

IDCs have a V-shape made of sharp-edged metal. The insulated wire is pushed down into the 'V', usually with a special tool, so that the sharp edges of the 'V' cut through the insulation material and bite into the copper conductor itself. There is normally some 'springiness' between the sides of the 'V' so that once displaced by the conductor they will permanently exert a force on the conductor, keeping it in place and maintaining a low attenuation connection. To be reliable the IDC is designed to work with a limited range of conductor sizes, typically 22–26 AWG, and not to be repeatedly used.

Some typical styles of IDC are the '110' (pronounced 'one-ten'), the LSA from Krone and the KATT from Mod-Tap/Molex. An earlier style, used in telephony but not datacoms, is the '66'. IDCs must be used with the correct tool specified by the manufacturer.

## 8.2.2   RJ45

The RJ45 plug and socket is the most widely used connector in copper cable based LAN equipment. RJ45 is a USOC (Universal Service Order Code) designation for specific applications and/or manufacturers, and the correct generic title is IEC 60603-7 (connectors for frequencies below 3 MHz for use with printed circuit boards

— part 7: Detail specification for connectors, 8-way, including fixed and free connectors with common mating features, with assessed quality) but RJ45 has come to be the commonly accepted term. Note that IEC 60603-7 specifies frequencies below 3 MHz and this standard is often seen more as a constructional specification, but Amendment no.1 of IEC 60603-7 will detail test methods and related requirements for use at frequencies up to 100 MHz. The electrical performance must be as detailed in ISO 11801 or TIA/EIA 568. Also often invoked is ISO/IEC 8877:1992 Information technology — telecommunications and information exchange between systems — interface connector and contact assignments for ISDN Basic Access Interface located at reference points S and T.

The RJ45 has eight pins so it can terminate a four pair cable. There are other 'RJ' styles such as the RJ11 which has six pins and is commonly used in telephony, and the RJ21x, which is a 50-pin, 25-pair connector.

LAN equipment usually comes with a row of RJ45 sockets. The main horizontal cable of the cabling system is terminated with another RJ45 socket that is housed in a patch panel. The LAN equipment and the patch panel are normally co-located within the same or adjacent equipment racks and are connected together with a patchlead of usually no more than three metres length. The patch cable has to have the same electrical performance as the main cable (but with slightly higher attenuation) and it is terminated with an RJ45 plug.

The RJ45 pins still have to be connected to the four-pair cable. This may be done by an IDC, which terminates the cable. The IDC is hard-wired to the pins of the RJ45 or it may be connected via a small printed circuit board, PCB, which helps to cancel out some of the crosstalk inherent in the connector design.

The RJ45 may be in unscreened or screened format and use any of the previously mentioned Insulation Displacement Connectors. Figure 8.3 shows an example.

Which conductor in the cable is connected to which pin on the front of the RJ45 is decreed in the standards (ISO 11801, TIA/EIA 568A and EN 50173) and is commonly referred to by its American designation, 568A and 568B. It is unfortunate that the names of the current and proposed overall cabling standard are also 568A and 568B.

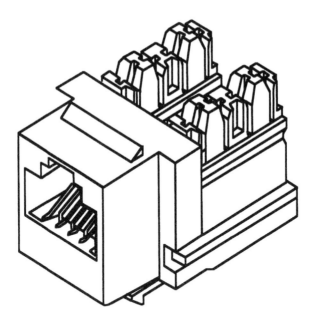

**Fig. 8.3** RJ45 plug and socket.

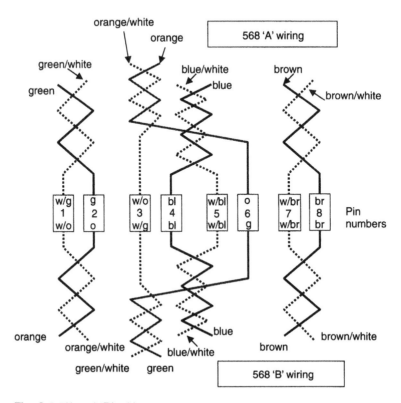

**Fig. 8.4** 'A' and 'B' wiring.

The two methods are often called 'A' or 'B' wiring. In essence, pairs one and three are swapped over as shown in Fig. 8.4. There is little, if any, technical difference between the two wiring standards but it is essential to remain consistent within the same installation.

## 8.2.3   Connector categories

The RJ45 was originally designed for humble 3 kHz telephone applications but it has been pressed into service for use up to 250 MHz. Clearly all RJ45s are not the same. Like cable, connectors also have categories, i.e. categories 3–7.

It is essential that all components in the cable system are designed to the same category of performance. The overall system performance of the cabling link is only as good as the lowest common denominator. For example, a cable system may have category 5 cable terminated with category 5 connectors, but if it is connected into the LAN equipment by category 3 patchleads then the overall channel performance will only be category 3.

## 8.2.4   Category 7

The RJ45 is unlikely to exceed the demands of category 6 with its requirement for a 250 MHz frequency range. Category 7 requires 600 MHz performance. ISO have approved a connector, see Fig. 8.5, designed by Alcatel which has category 7 performance yet is backwardly compatible with RJ45 connectors. It achieves this by having eight pins in the socket, as in a conventional RJ45, and four more, arranged as two pairs in each corner. When in RJ45 mode, i.e. category 6 and below, only the 8 pins on the top are engaged. When the socket detects category 7 operation is required (by detecting a

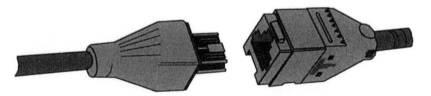

**Fig. 8.5** Category 7 connector.

lug on the side of a category 7 plug) the four middle pins on the top deck are shorted out and the four pins on the lower side are engaged, thus giving sufficient physical separation of the four pairs so that category 7 performance is achieved.

As an alternative design, the ISO committee has approved a connector design from The Siemon Company for category 7.

# 8.3   Patch panels

The RJ45 sockets are organised in a patch panel. Patch panels are 19 inches wide (483 mm) to fit into the industry standard equipment racks. Perversely the units for all three dimensions of a patch panel are different. The width is always 19 inches, the depth is in millimetres and the height is in 'U'. One U is 44 mm. The standardised notation of U allows designers to quickly calculate how much equipment can be loaded into an equipment rack. So for example a 30 U rack could accommodate 30 × 1 U patch panels. A 1-U patchpanel, see Fig. 8.6, will usually have 12, 16 or 24 RJ45 sockets on the front. An IDC on the back will terminate each cable and supply the connection through to the pins of the RJ45 on the front. There may or may not be extra cable management supplied on the back. The patch

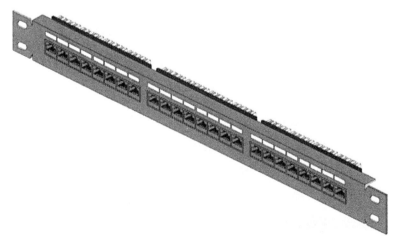

**Fig. 8.6** Patch panel.

panel may be screened or unscreened and wired in either 568A or 568B format. A patch panel may be larger than 1 U. A 2-U panel can typically terminate up to 48 connectors and there are even 4-U panels for 96 connectors. The RJ45 patch panel can appear in any of the areas described as distributors or cross-connects.

## 8.3.1 Cross-connect strips

Strips of insulation displacement connectors (IDCs), typically 50 or 25 pairs in a row are often called cross-connect or cross-connect strips. The concept of cross-connect is usually taken to mean one set of connectors connected by patchlead to the LAN equipment and another set of connectors connected to the main cabling. Jumpers join the two sets of connectors. Figure 8.7 demonstrates this. Two RJ45 patch panels could also do the job of cross-connect, but the term is usually applied to strips of IDCs. The IDCs can be strips of the 110 or LSA style of connectors or some other proprietary makes (see Fig. 8.8).

The IDC in a cross-connect usually works by having a 'lower deck' which is just a row of 50 IDCs. Onto this is placed a connector block that is made for two, three, four or five pairs. The exit wire, or jumper

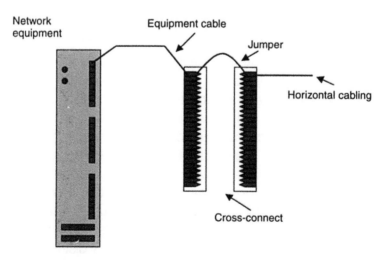

**Fig. 8.7** Cross-connect method.

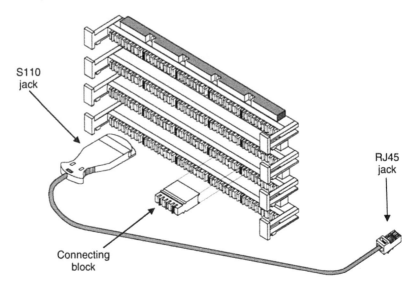

S110
jack

RJ45
jack

Connecting
block

**Fig. 8.8** 110 Cross-connect.

can be punched-down onto this upper deck formed by the addition of the connector block. There are special connectors that can fit directly onto the top of the IDC to provide an easily removable jumper lead.

The cross-connect is also used to break down high pair-count backbone cables into smaller pair-count cables. For example a one-hundred pair backbone may need to be broken down into four-pair cables for onward routing to the desk position. In such a case twenty-five pairs at a time would be laid down onto the lower deck of the IDC cross-connect. A four-pair connector block is pushed down onto the first four terminated pairs of the backbone cable and makes electrical contact with them. The four-pair exit cable is then punched down onto the top-deck of the connector block. In telephony terms such an arrangement would be called an IDF or intermediate distribution frame. The cross-connect arrangement can appear in any of the areas described as distributors or cross-connects.

## 8.3.2  Patchleads

A patchlead is a short length of stranded conductor cable with an RJ45 plug on each end. It links patch panels, cross-connects and the active LAN equipment at each end of the cabling channel. The exact terminology depends upon the supplier and where the patch cable is used, e.g.

*Generic terminology*:

- Patchlead.
- Patchcord.
- Patchcable.

*Topographic terminology*:

- Work area cables: connect telecommunications outlets to terminal equipment.
- Equipment cables: connect IT equipment to the generic cabling at distributors.
- Patch cables: used at a cross or interconnect.
- Jumper: an assembly of twisted pairs without connectors used at a cross-connect.

The latest standards are using the word 'cord' to describe any user-moveable cable.

# 8.4  Telecommunications outlets

A telecommunications outlet (TO) is the connector interface that the user is going to encounter in his/her place of work. The TO will house at least two connectors and may be floor or wall mounted, see Fig. 8.9. It may even be incorporated into the desk furniture. The TO will contain at least one RJ45. The other connector can be another RJ45 or an ST or SC optical connector. The density of TOs in the workplace should be at least one per individual work area and at least one per 10 square metres. The TO should preferably be within three metres of the user's workstation.

The TO will normally be of a plastic construction that fits in with

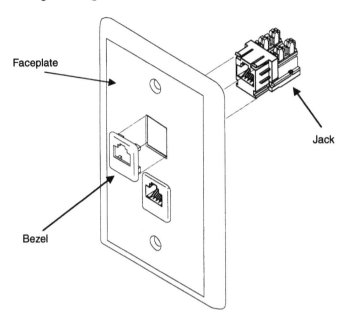

**Fig. 8.9** Typical telecommunications outlets.

the existing decor, national standards and cable containment system in use. In many countries it is expected that a spring-loaded shutter will be provided to keep dust and foreign objects out of the connector socket.

# 8.5  Consolidation points

Three types of junction points are allowed in the standards. They are:

- transition point
- consolidation point
- MUTOA (multi-user telecommunications outlet assembly).

## 8.5.1  Transition point

A transition point forms a connection in the horizontal cable between the TO and the Floor Distributor/Horizontal Cross Connect. Only one is allowed and it must be seen as semi-permanent, i.e. not a *point of administration* as the standards put it. The transition point may incorporate any of the RJ45 or IDC interconnection methods previously discussed. It should be noted in the American Standard, TIA/EIA 568, that a transition point specifically means the interface between a conventional horizontal cable and a flat, undercarpet cable.

## 8.5.2   Consolidation point

A consolidation point (CP) may be used in open plan offices to connect between a number of horizontal cables from the floor distributor and telecommunications outlets. This may be useful in an open plan office environment where TOs may have to be moved. It could also be used in a zoned cabling system for instance where a twenty-five pair category 5 cable is connected to a 'zone' and six four-pair cables then proceed onto the work areas. A CP should not serve more than twelve work areas and as it may be a point of administration it may look quite like a conventional patchpanel. TIA/EIA 568 defines it more specifically as an interconnection between horizontal cables in the building pathways to horizontal cables extending into furniture pathways.

## 8.5.3   MUTOA

The MUTOA is another point of administration serving up to twelve work areas but this time the cables extending to the user are stranded patchleads, so there is no further termination in a TO. The concept is also aimed at serving open plan office areas where frequent moves and changes may be required. The MUTOA was first seen in TSB75 where longer patchleads might be used at the expense of shorter fixed horizontal cable. The standard model allows for 90 m of fixed cable with a combined total of not more than 10 m of patchlead, so the channel maximum length is 100 m. TSB75 modifies this to allow up to 27 m of patchlead with the final work area cable taking up to 20 m of that. But the main horizontal cable would have to be reduced to 70 m under these conditions. The open plan cabling concept will be incorporated into both TIA/EIA 568B and ISO 11801 2nd edition 2001. Note that the American model presumes the use of 24 AWG patchlead, which has an attenuation of 1.2 times the standard solid cable. ISO and EN standards allow the use of 26 AWG cable because the patchlead may have up to 50% more attenuation than the fixed cable; thus the length allowances would have to be proportionally reduced if 26 AWG cable was used.

# 8.6   Other devices

## 8.6.1   Baluns

A balun (balanced — unbalanced) is an impedance matching device that can be used to connect an unbalanced coaxial cable of impedance between 50 and 125 Ω and a balanced 100 Ω twisted pair cable. The balun has to be specifically made for the application so one might be described as 75–100 Ω for example. The 100 Ω side would be a standard 8-pin RJ45 and the 75 Ω side would be the appropriate BNC coax connector. Baluns may be screened or unscreened and sometimes there can be different pins connected within the RJ45, so they are best bought from the same supplier, in pairs. The electronic transmission equipment, which may have been made to work on 93 Ω coax, should be unaware that it is really communicating over 100 Ω four-pair cable, if the correct baluns have been used.

Video baluns can be baseband (8 MHz composite video, e.g. a CCTV camera), broadband (550 MHz CATV) or RGB, red, green and blue signals on three separate cables for high quality graphics.

## 8.6.2   Adapters

Devices such as telephones and ISDN equipment can work over structured cabling but usually cannot plug directly into the RJ45 telecommunications outlet. Some form of adapter is often required which is often dictated by the style of PABX in use. The manufacturer of the PABX/ISDN equipment should be consulted to determine which style of adapter is required, e.g. secondary line adapter, PABX line adapter or PSTN line adapter.

Other devices include media filters, which connect the 150 Ω DB9 connector of a PC Token Ring card to 100 Ω structured cabling, and video baluns.

It must be noted that the standards only allow for RJ45 copper cable connectors or ST or SC optical connectors in the TO, nothing else is allowed. So any balun or adapter must be external to the RJ45. This allows the device to be removed at any time letting the cabling revert to its open-systems, applications-independent role.

## 8.6.3   Cable organisers

Racks containing large amounts of patch panels can create unmanageable piles of patchleads cascading down the front of them. A 30 U rack could have 720 patchleads of an average three-metre length in the front of it, i.e. over 2 km of cable! Cable managers and organisers should be inserted in between patch panels at a spacing of every 3 or 4 U to route the patchleads neatly away from the front of the patchfield.

## 8.6.4   Line protection devices

Where external copper cables are used they should be protected from over voltage and surge current conditions at the point where they enter the building. Lightning strikes in particular can cause damaging and dangerous voltage spikes to enter the building cabling, posing a threat to life and equipment.

Primary protectors should be placed across the incoming circuit and ground as close as possible to where the cable enters the building. For service entrances this may be the point of demarcation between the PTT cable and the user's own cabling. There are three main types of primary protector:

- Carbon block.
- Gas tube.
- Solid state.

All work on the principle that at a certain voltage the route to earth will break down and shunt the fault current to ground. There are also devices known as secondary protectors which are often designed to defend against lower level, but perhaps more persistent fault or 'sneak-currents'.

Any device attached to a high speed LAN cable must be designed so that it does not degrade the high performance potential of the cable.

# 9

# Copper cable technology — transmission characteristics

## 9.1 Introduction

In this section all the principal transmission characteristics of a copper cable are defined along with some explanations of why they are important. Some of these parameters and their relevance to real life communications performance are also touched upon in Chapter 3. All of the copper cable and system parameters quoted in EN 50173, ISO 11801 and TIA/EIA 568 are defined here.

It must also be noted that whenever specifications are quoted they nearly always mean when measured at 20°C. Cable performance always declines as the temperature goes up. The IEC believes that the temperature attenuation coefficient for any UTP is 0.4% per °C and when there is at least one screen between the insulation and the jacket it is 0.2% per °C. They are also considering the effect of temperature on ELFEXT and unbalance attenuation.

## 9.2 The fundamental units of measurement

- Capacitance, C, farads (pF).
- Inductance, L, henries (μH).

- Resistance, R, ohms (mΩ).
- Conductance, G, siemens (mS).

*Note*: A definition of these units is given in Table 3.1.

# 9.3   The cable as a transmission line

A length of cable acts as a communications channel for information transmitted down it. The performance and behaviour of the signal is defined by the fundamental parameters of the cable (as well as external noise) and the channel is described as a transmission line. Note this term is also used in electrical engineering to describe the performance of high voltage cables as well.

Figure 9.1 shows the cable as 'T-circuit' transmission line. The cable has a resistance, $R$, an inductance, $L$, a capacitance, $C$, and a conductance, $G$. The latter represents the loss within the insulation material which can be represented as a resistance leaking energy from one conductor to another. The inverse unit of ohms is used, i.e. siemens.

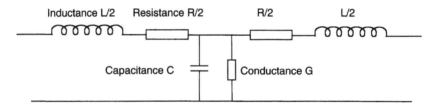

**Fig. 9.1** A two-conductor cable depicted as a transmission line (equivalent 'T' circuit) with inductance, resistance, capacitance and conductance.

## Impedance and characteristic impedance

Impedance $(Z)$ is the opposition of a circuit to an alternating current caused by the capacitance, resistance and inductance of the circuit. Characteristic impedance $(Z_0)$ is the impedance presented by an infinitely long length of a uniform transmission line. A cable terminated with a load equal to the characteristic impedance will transfer all the

**Fig. 9.2** Attenuation down a length of cable.

energy to the load without any back reflections. In standard data cable the characteristic impedance is $100\,\Omega$. The transmitter and the receiver connected to that cable must also have impedances of $100\,\Omega$ or some of the energy will be reflected back. This can be extremely disruptive to a high-speed data circuit (see Return loss pp 140–1).

## Attenuation

Attenuation is the amount of energy absorbed by the transmission medium as the signal travels down it, as shown in Fig. 9.2. Attenuation can be in decibels if quoting the loss of a discrete channel or component but is more commonly given as a loss per unit length, such as dB/km, or dB/m or even dB/100 m.

If measuring voltages then the attenuation is $20\log V_1/V_2$, where $V_1$ is the voltage measured at the input end of the cable and $V_2$ is the voltage measured at the receiving end. As the voltage at the far end will be less, due to the attenuation, then the resulting figure will be negative, e.g. $-20\,$dB when expressed in decibels.

If measuring power then the attenuation is $10\log P_1/P_2$. Sometimes the term *Insertion loss* is used. It is synonymous with attenuation but may be used to express the effect on attenuation of a component, such as a connector, when inserted into a circuit.

## 9.4   Cross talk and power sum measurements

- Near end cross talk, NEXT.
- Power sum cross talk, PSNEXT.
- Far end cross talk, FEXT.

- Equal level far end cross talk, ELFEXT.
- Power sum equal level far end cross talk, PSELFEXT.
- Alien cross talk, ACT or AXT.

Cross talk is the process of the signal in one circuit, or pair of wires, being picked up by an adjacent circuit, or pair of wires. Cross talk is the greatest source of noise or interference on a twisted pair cable, and is much larger than any externally generated interference. It is essential that cross talk is kept under control to achieve successful high-speed transmission of data on a circuit.

The principal design weapon to combat cross talk is to improve the balance of the circuit, so that an equal and opposite current flows down each wire in a pair. Each pair in a cable is twisted to reduce the magnitude of the electromagnetic field around it, and to reduce the circuit's susceptibility to the electromagnetic field from other nearby pairs. The amount of twist, or lay length as it is called, is different for each of the four pairs in a standard data cable to reduce the pick-up effect of long parallel runs, which could be up to 90 m in a structured cabling system. There are different types of cross talk.

## 9.4.1 NEXT

NEXT is the transfer of energy from one pair to another pair when measured right next to the transmitter (see Fig. 9.3).

$$NEXT = 20\log V_1/V_2 \text{ dB},\qquad\qquad [9.1]$$

where $V_1$ is the input signal voltage measured next to the transmitter, and $V_2$ is the induced voltage measured on any adjacent pair.

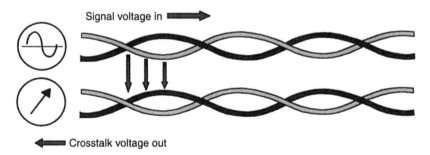

Signal voltage in ➡

Crosstalk voltage out ⬅

**Fig. 9.3** Near end cross talk.

## 9.4.2  PSNEXT

In a conventional data transmission system it is common to find one pair in a cable dedicated to the transmit circuit and another pair dedicated to the receive circuit. Ten megabit/s Ethernet, 10baseT, is a typical example. NEXT performance is important to see how each pair will interfere with the other. In more modern protocols however there is a move to maximise the bandwidth potential of the cable by spreading the signal over all four pairs of a LAN cable, for example, 100baseT4 and the gigabit Ethernet system 1000baseT. Because of these applications, and doubtless more to follow in the future, it is more realistic to consider the effect of how every pair interacts with every other pair in the cable. In a four pair cable, any one pair is subject to interfering signals from all three of the other pairs simultaneously, if the other three pairs are all energised by signals (see Fig. 9.4).

Similarly in a 25-pair cable, any pair is subject to interference from up to 24 other pairs simultaneously. The other pairs could all be carrying different signalling protocols, analogue or digital or even telephone ringing signals.

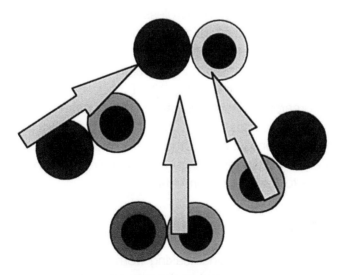

**Fig. 9.4** Power sum crosstalk is the combined effect of every other pair in the cable acting simultaneously upon the measured pair.

Adding up all these effects gives rise to the term, power sum measurements. The calculation is essentially to take the anti-log of all the NEXT figures, add them up and take the logarithm of the result to return a PSNEXT figure in decibels. If we take the worst case example, whereby three adjacent pairs are inducing a NEXT voltage into the fourth pair, all of the same magnitude and all adding in phase, we can see the worst case result of power sum crosstalk:

For example, if the pair-to-pair NEXT of three cables is 27 dB, the resulting power sum calculation is:

$$\text{PSNEXT} = 10\log\,(\text{antilog}\,(-27/10) + \text{antilog}\,(-27/10)$$
$$+ \text{antilog}\,(-27/10))$$
$$= -22.2\text{dB, or a 4.77dB reduction in cross}$$
$$\text{talk isolation.}$$

Astute mathematicians will see that 4.77 is just ten times the logarithm of three, so the reduction in cross talk isolation would always be 4.77 dB, whatever the starting NEXT figure was, as long, of course, that all the pair-to-pair NEXT signals were the same and they added up in phase. In reality this is not the case, and a more practical effect is to reduce the worst case pair-to-pair NEXT value by approximately 3 dB to arrive at the PSNEXT figure.

The method is the same regardless of how many pairs are in a cable, but once we have a cable with more than seven pairs then the pair-to-pair NEXT values will be very small for pairs separated by other pairs. The PSNEXT will not change greatly for larger cables. There are few better ways of improving NEXT performance than by just obtaining a greater physical separation of the pairs.

Twenty-five pair cables have to be laboriously tested, pair by pair, and the combinations all added together, preferably by computer, to arrive at the PSNEXT performance of every pair; a test that will never be conducive to field measurements.

## 9.4.3   FEXT

FEXT is the amount of signal leaking from the transmit pair to the adjacent pair when measured at the other end of the cable. It used to be assumed that NEXT and FEXT performance would be the same, but practical measurements have shown this not to be the case. FEXT has become important because of protocols such as 1000baseT utilising simultaneous bi-directional transmission down each pair, and the fact that FEXT is much harder to cancel out electronically than NEXT.

With a large amount of cheap digital signal processing power at the disposal of the hardware designers, a lot of the transmission problems associated with cables can to some effect be cancelled out by sophisticated signal processing. Take for example delay skew, whereby signals arrive at different times at the far end of the cable due to the characteristics of the cable. If the receiver circuitry can assume that all the signals started off at the same time, then a simple test message sent at the start of a transmission session will allow the receiver to calculate the delay imposed by the cable channel on the four separate signals. All further signals can be compensated for by the receiver delaying the first, second and third fastest signals by the appropriate amount, in digital registers, so that when the fourth, and slowest, packet of data arrives it can be fed simultaneously with the other three packets of data onto the next stage of the receiver circuit.

Similarly the transmitter circuitry can accommodate a certain amount of NEXT. If each transmitter sends a known signal in turn onto each of the pairs to which it is connected, and, at the same time the receiver circuit connected to the three adjacent pairs 'listens' to any signal appearing at the same time on those adjacent pairs, then they can reasonably assume that they are detecting the crosstalk between those pairs. If, for future transmissions, a suitable 'inverse' signal is applied to the other receiving pairs then a large part of the NEXT can be electronically cancelled out.

FEXT is much harder to accommodate. For a start the effect is appearing about 500 ns later at the far end of the cable and is mixed with other crosstalk and external noise components. Also what effect will any compensation for FEXT have on the NEXT performance?

Thus FEXT, along with external noise, is extremely difficult to cancel out by digital signal processing, and so it is essential to design the cable system with these effects minimised (see Fig. 9.5).

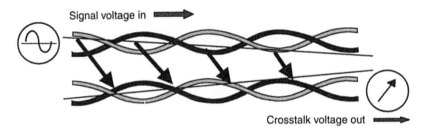

Signal voltage in ➡

Crosstalk voltage out ➡

**Fig. 9.5** Far end crosstalk, FEXT.

### 9.4.4   ELFEXT

The effect of FEXT can be masked by the presence of the attenuation of the cable. The latest structured cabling standards thus require the ELFEXT to be measured and reported, that is, equal level far end crosstalk. ELFEXT is the difference between the FEXT and the attenuation of that link. It is similar to attenuation to cross talk ratio (ACR) but viewed from the other end of the cable. By putting the attenuation of the cable back into the equation it gives a length-independent, normalised value which is of more use than FEXT on its own.

### 9.4.5   PSELFEXT

As with PSNEXT, which is a more representative function of the overall near end crosstalk of a cable, ELFEXT is complemented by PSELFEXT. PSELFEXT is the sum of the combined effect of all the FEXT of all the other pairs in the cable, less the attenuation of the victim pair. With NEXT, in a four pair cable, there will be six possible combinations. Power sum measurements will have four combinations. It is usual for the channel to have all these parameters measured from both ends, and then the worst case pair combination for each parameter to be reported.

### 9.4.6   ACR

The ACR (attenuation to cross talk ratio) is similar in concept, although not identical, to signal to noise ratio (SNR). It is a good measure of

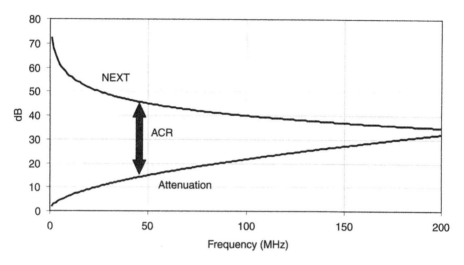

**Fig. 9.6** Attenuation to crosstalk ratio.

the overall quality of a cabling link. The higher the ACR the better as it implies that the desired signal is not being so severely attenuated that the effect of cross talk noise will become too significant. In communications channels it is generally considered that a positive value of ACR is required for successful, error free transmission.

Figure 9.6 shows how attenuation increases as the frequency increases. The near end crosstalk separation also declines as the frequency increases. The gap between the two functions, the ACR in decibels, thus gets less and less as the frequency increases.

ACR = NEXT – attenuation                                                                 [9.2]

PSACR is the PSNEXT value of the victim pair minus the attenuation of that pair.

## 9.5  Nominal velocity of propagation

The speed that the signal travels down the copper cable compared to the speed of light in a vacuum is called the nominal velocity of propagation, or NVP. For most data cables it is in the range 0.69–0.75. If the speed of light is 300 000 km/second and if the NVP of a cable is 0.7 then the speed of the signal in the cable is 0.7 × 300000 or 210000 km/s. The NVP is related to the dielectric constant by the following formula:

$$NVP = \frac{1}{\sqrt{\text{dielectric constant}}} \qquad [9.3]$$

The value of the NVP is an important factor to know to load into test instruments so that the correct length of cable can be determined.

### 9.5.1   Delay, differential delay and asymmetric skew

The length of time taken for the signal to transit the entire cable system is known as the delay. The latest standards call for a delay of no more than around 570 ns to transit the entire length of a 100 m cabling system. Some LANs such as Ethernet are sensitive to time delay in the transmission medium.

A four pair cable consists of four communications channels, each having a slightly different delay time. The difference in delay is caused by the differing lay length on each pair (thus giving a different physical length of each pair) and possibly by the use of different insulation materials between the pairs. LAN protocols that share the data across all four pairs such as 100baseT4 and 1000baseT are very sensitive to differential delay. The terms used may be skew, asymmetric skew or differential delay. The standards call for no more than 45–50 ns delay between the 'fastest' and 'slowest' pair in a cable system.

## 9.6   Return loss

Return loss (RL) is the amount of energy reflected back from a circuit due to an impedance mismatch between the source, the cable, the load, or all three. If the input impedance of the source matches the characteristic impedance of the cable, which in turn is terminated by a load of the same impedance, then all the signal energy, minus that lost due to attenuation, will be transferred to the load. Failure to do this will represent an inefficient loss of energy. But more importantly,

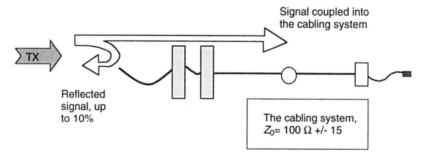

**Fig. 9.7** Return loss.

out-of-phase signals reflected back into the transmitter could create havoc with modern multi-level, multi-phase coding schemes.

Load matching would be ideal but is not practical in the real world. For example the $100\,\Omega$ cable normally specified in LANs is in reality allowed to be $100 \pm 15\,\Omega$ and still be within specification.

Instead of measuring the characteristic or input impedance of the circuit, which is difficult, it is considered more practical, and useful, to measure the result of impedance mismatches, i.e. return loss. Between 3 and 10% (according to frequency) is allowed to be reflected back in modern cabling systems (see Fig. 9.7).

## 9.7   TOC and common mode and differential mode transmission

The purpose of a twisted pair balanced cable is that an equal and opposite current flows in each conductor. This is called differential mode transmission. An effect of external noise is to cause a common mode current to flow in the same direction along each conductor, with the return path along the ground plane. This can have a severe effect upon the noise performance, and rejection, of the cable system. An often-used remedy for this is a transformer coupled receiver circuit with a centre tap to earth. The common mode noise current can then be cancelled out through the centre tap. Figures 9.8, 9.9 and 9.10 attempt to demonstrate this.

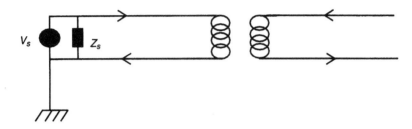

**Fig. 9.8** Differential mode current.

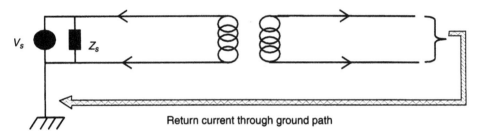

Return current through ground path

**Fig. 9.9** Common mode current.

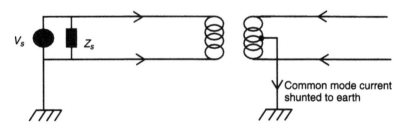

Common mode current
shunted to earth

**Fig. 9.10** Common mode termination. A centre tap on the receiver transformer can shunt much of the common mode noise current to ground.

The NEXT and FEXT performance of a circuit is not exactly the same when it is configured in the differential mode and common mode terminations, that is, with or without an earthed centre tap on the receiver transformer. NEXT figures for components should therefore be reported as the worst case when measured in circuits terminated in both configurations.

## 9.7.1   TOC and the effect of differing NEXT performance of components

A connector and its accompanying socket have to be matched to some extent to achieve a worthwhile combined NEXT performance at high frequencies when mated together. It is a misconception that putting a high performance plug into a lower performance socket, e.g. a patch cord into a patch panel, will raise the overall performance. This is not the case. The combined NEXT performance will be worse than for the two mediocre components together. It is a measure of the quality of the components if they can tolerate a wide difference in performance and still return an overall acceptable performance. Such measurements must be made in the common mode and differential mode termination to demonstrate an acceptable performance. Figure 9.11 shows from a practical experiment the effect on the overall NEXT of a circuit when patchcords with differing individual NEXT performances are plugged into it.

Figure 9.11 also shows the effect on three different patchpanels when different patchleads of varying NEXT performance are plugged

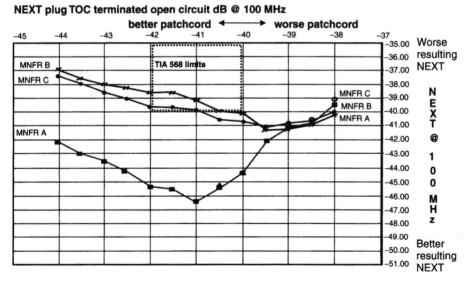

**Fig. 9.11** NEXT performance of a patchpanel with differing quality patchcords inserted into it.

into them. The patchleads are open circuit at the other end, hence the name terminated open circuit (TOC).

A 45 dB patchcord is very good and a 37 dB patchcord is very poor. The American TIA 568A standard requires that the resulting NEXT performance of the patchpanel should be better than 40 dB when patchcords of 40–42 dB NEXT are plugged into them. This TOC test should report the worst case performance when the test is done with common mode and differential mode terminations. Perhaps surprisingly the overall NEXT performance declines even though better and better quality patchcords are being added to the system, with peak system performance being obtained for the three manufacturers for patchcords with NEXT performances between 39 and 40 dB. The test is done on the worst performing pair combinations and all at 100 MHz.

# 9.8   Electromagnetic emissions and immunity of cable

The concept of a twisted pair cable incorporates the intention to reduce electromagnetic emissions (sometimes referred to as electromagnetic compatibility or EMC) by having an equal and opposite current flowing down each conductor of the cable pair, as seen in Fig. 9.12. Thus the electromagnetic field created around the cable as a result of these currents is in opposition to each other and will tend to cancel each other out. Similarly any external noise induced into the cable will cause the induced current to flow in the same direction down each conductor (common mode noise) and so the receiver

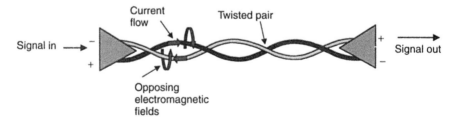

**Fig. 9.12** Electromagnetic emissions from a twisted pair cable.

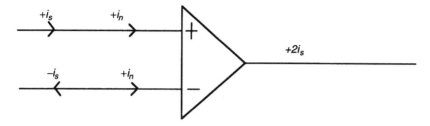

**Fig. 9.13** A differential receiver can cancel out some of the induced noise.

differential circuitry, Fig. 9.13, can cancel most of it out. If the desired signal current flowing down one conductor is $i_s$, then the signal on the other conductor is $-i_s$. The noise current induced on each conductor will be $i_n$.

Equation 9.4 shows that the noise signal is cancelled at the output as the receiver subtracts one current from another:

$$(i_s + i_n) - (-i_s + i_n) = 2i_s. \tag{9.4}$$

The immunity of a device or circuit to electromagnetic interference is known as electromagnetic immunity (EMI).

The capability to reduce emissions and reject noise depends upon exactly the same current flowing down each conductor and the same voltage appearing between each conductor and ground at any adjacent point along the two conductors of the pair. This is a function of the cable balance which in turn depends upon the accuracy and consistency in the construction of the cable, such as core concentricity, conductor diameter and insulation thickness.

The number of twists in the pair (also known as lay length; the shorter the lay length equates to more twists in a pair) also reduces the distance from the cable at which the cancelling out effect of the two opposing electromagnetic fields takes place. For numerous pairs within an overall cable sheath this effect needs to take place as close as possible, so generally the more twists in the cable the better. There is a practical limit to this of course and the shorter the lay length then the more expensive the cable will become.

### 9.4.1   Near end unbalance attenuation

A way to measure the cable balance is to consider the near end unbalance attenuation (NEUA), which is proposed in prEN 50288. If the cable is not perfectly made then the two currents will not be exactly equal and the two voltages across the resistors will not be equal either. NEUA replaces two earlier proposed measurements, longitudinal conversion loss (LCL) and longitudinal conversion transfer loss (LCTL). NEUA is defined in equation 9.5 and the test methodology demonstrated in Fig. 9.14.

$$NEUA = -20\log\frac{V_3 - V_2}{V_1}. \tag{9.5}$$

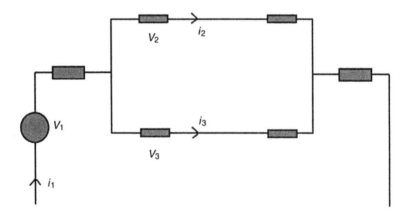

**Fig. 9.14** Measuring near end unbalance attenuation.

### 9.4.2   Coupling attenuation

Another way of deciding upon the quality of cable balance is to study the practical effect of imbalance by measuring the radiated field from a cable or by measuring the induced current in the cable. This is done by the so-called absorbing clamp method. Radiation and immunity are reciprocal. What makes a good radiator will also make a good receiver. Table 9.1 gives a proposed method from 3P Laboratories of Denmark to give an EMC rating for cables and components. Figure 9.15 demonstrates the method for measuring coupling attenuation.

Table 9.1 Characteristic ranges of EMC performance and coupling attenuation

| Rating | Cable Coupling attenuation (dB) | Component Coupling attenuation (dB) | Characteristic |
|---|---|---|---|
| 1 | 1–10 | 1–4 | Very poor |
| 2 | 11–20 | 5–14 | Very poor |
| 3 | 21–30 | 15–24 | Very badly connected screens. Poorly balanced UTP |
| 4 | 31–40 | 25–34 | Badly connected screens. Poorly balanced UTP |
| 5 | 41–50 | 35–44 | Min. requirement for 100 MHz UTP cables |
| 6 | 51–60 | 45–54 | Low quality FTP |
| 7 | 61–70 | 55–64 | Medium quality FTP, min. spec for 200 MHz screened cables |
| 8 | 71–80 | 65–74 | Fine quality FTP, low quality S-FTP |
| 9 | 81–90 | 75–84 | Min. requirement for 600 MHz cables |
| 10 | Min. 91 | Min. 85 | Especially screened cabling |

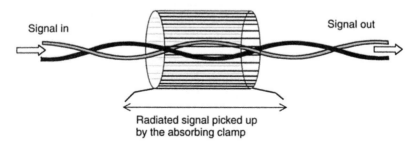

Signal in

Signal out

Radiated signal picked up
by the absorbing clamp

**Fig. 9.15** Measuring coupling attenuation.

Parameters such as coupling attenuation (CA) demonstrate the end effect of the cable's efficacy at keeping in or keeping out electromagnetic signals by literally measuring what is radiated or absorbed. It is possible however to measure the efficiency of a cable screen directly. Surface transfer impedance (STI) is the method described for signals below 30 MHz. A current is circulated through the screen and the induced voltage across the conductors, which are

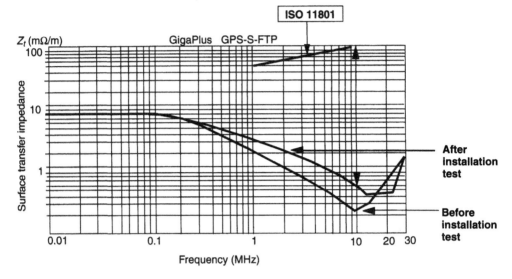

**Fig. 9.16** Surface transfer impedance.

all tied together, is measured. A length of the cable sits within a brass tube that is also connected to the screen. With a current and a voltage measurement we can arrive at an impedance measurement, the units are in milli-ohms per metre.

For measurements above 30 MHz, the parameter is known as screening efficiency ($SE_f$) and the units are decibels. Figure 9.16 shows surface transfer impedance against frequency for a screened cable. Note how the performance deteriorates after the simulated installation test.

# 10

# Copper cable technology — testing

## 10.1 Introduction

Testing may be split into two parts:

- laboratory testing to type approve components and systems
- field, or on-site or acceptance testing.

Manufacturers and third-party approval houses do the former. The latter is done by the installer to prove to the end-user that what he has installed is up to specification. The manufacturer of the cabling system may also want to see these test results before issuing a manufacturer's warranty for the site.

Laboratory testing is done by expensive and sophisticated machines called network analysers that require skilled operators. For acceptance testing there is a range of simpler hand-held test equipment. Such equipment may be very simple, such as a continuity tester. There does exist today however a class of testers specifically designed to automatically test a structured cabling system for all the parameters described in the standards. These hand-held testers are a mixture of voltage meter, time domain reflectometers and microprocessors with the capability to store hundreds of test results and sometimes display them graphically as well. They incorporate remote injectors that plug into the far end of the network enabling all parameters to be tested and automatically fulfilling the standards requirement of testing from both ends.

The standards require that the worst case value or margin is reported from the two ended tests across the frequency range of interest, e.g. 100 MHz for category 5 and 250 MHz for category 6. At the time of writing there were no hand-held testers capable of measuring the 600 MHz range of category 7. Some leading testers with category 6 capability are Microtest Omniscanner, Fluke DSP 4000, Wavetek 8600, HP Scope 350, Datacom Technologies LANcat 6.

Modern hand-held testers now usually incorporate a facility for adding on a fibre optic test module.

## 10.2   Time and frequency domain measurements

One of the most powerful features of many testers is time domain reflectometry (TDR). TDR involves launching a signal into the transmission media and seeing if any impedance discontinuities reflect the signal back to the source. Such discontinuities could be open circuits, such as the end of the cable, or even short circuit faults. Any major deviation from the expected characteristic impedance will cause a reflection, such as a connector or splice anywhere in the cable.

Having launched a signal, and timing when its reflection returns, then the microprocessor, having been told at what speed the signal travels, can calculate the length of the cable and the relative position of any other major discontinuity in the circuit. The time information can be translated into distance information on the horizontal axis of the display because the processor knows the speed of the signal.

The tester can only know the speed of the signal if it is told, and it receives this information when the operator inputs the (correct) value of NVP for the cable. The NVP (nominal velocity of propagation) is the ratio of the speed of the signal in the cable compared to the speed of light in a vacuum. So for a typical cable NVP of 70% the processor can determine that the signal must be travelling at 0.7 × 300 000 or 210 000 km/s. Incorrect values of NVP, or simply allowing the tester to go to a default setting will result in incorrect length calculations.

The above TDR principle is exactly the same as is used in radar and the optical fibre equivalent, the OTDR (optical time domain reflectometer). Copper cable based TDRs can also suffer the same problems as radar and OTDRs. The problem relates to the pulse length used and the subsequent 'dead-zone'. If the TDR sends out a pulse length thirty nanoseconds long then the 'length' of that pulse on the cable is going to be about 6.3 m (NVP 0.7, speed, 210 000 000 m/s). So the back end, or trailing edge, of the pulse will not leave the TDR until the front end, or leading edge, is 6.3 m down the cable. The receiving circuitry cannot turn on until the pulse has completely left the transmitter. So anything happening on the cable within the first 6 m will not be seen by the TDR. This is the dead-zone. Obviously as short a pulse length as possible would be best. This is offset by the cost and complexity of creating and managing very short pulses and the small amount of power they transfer into the circuit. The power transfer is not really an issue in cable testers designed for structured cabling systems because they only ever expect to look at a few hundred metres of cable at most (the standards require a capability to measure at least 310 m, roughly 1000 feet). OTDRs on the other hand might have to look at 100 m or 100 km of cable. The short pulse required to look at 100 m would be totally lost over 100 km. OTDRs therefore have the option of setting the pulse length depending upon the distance observed and the accuracy required. The shorter the pulse length then the more accurate the distance and event readings will be. If the pulse length is 6 m then no event can be identified to an accuracy better than plus or minus 3 m and events closer than 6 m apart could not be seen as separate items.

Fourier transforms allow the principal measurements to be made in the frequency domain rather than the time domain (see chapter 5 for a fuller account of Fourier transforms). Some testers measure in the time domain (i.e. a time domain reflectometer) and use Fourier analysis to convert into the frequency domain for frequency related parameters such as NEXT and attenuation. Other testers measure in the frequency domain and use Fourier transforms to arrive at time related phenomena such as cable length or distance where a possible fault may be occurring on the cable. Knowledge of the cable's NVP, which relates to velocity of the signal down the cable,

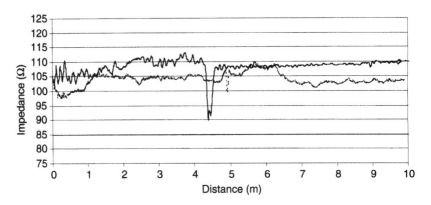

**Fig. 10.1** Frequency and time domain results for FEXT in a cable system.

allows the time axis on the time domain analysis to be converted to a distance measurement. Figure 10.1 shows an example of frequency domain and time domain cable measurement.

## 10.3   Test parameters

Table 10.1 summarises the required tests published in current and draft standards. The standards referred to are explained in more detail in chapter 15 and appendix I.

Table 10.2 gives the proposed test schedule from ISO 11801 2nd edition. It lists reference tests, i.e. normally done by the manufacturer and acceptance tests to be done on site by the installer.

Tables 10.1 and 10.2 are in close but not exact agreement. Note that ACR and PSACR are not included anywhere for testing even though they are specified in many standards. A specification for a project would be well advised to list all the parameters of Table 10.1 and add ACR and PSACR for good measure. Most automatic testers have default programmes based on only one standard and so would miss out some of the relevant parameters, though it is usual to allow customised test programmes to be built up by the installer.

Table 10.1 Test requirements from different standards

| | ANSI/TIA/EIA 568A | | | TIA/EIA 568B | IEC 61935 |
|---|---|---|---|---|---|
| | TSB 67 | TSB 95 | Addendum 5 | | |
| Wire map | x | x | x | x | x |
| Length | x | x | x | x | |
| Attenuation | x | x | x | x | x |
| NEXT pair to pair | x | x | x | x | x |
| NEXT powersum | | | x | x | x |
| ELFEXT pair to pair | | x | x | x | x |
| ELFEXT powersum | | x | x | x | x |
| Return loss | | x | x | x | x |
| Propagation delay | | x | x | x | x |
| Delay skew | | x | x | x | x |
| DC loop resistance | | | | | x |

Table 10.2 Acceptance and reference standards from ISO11801: draft (2nd edn)

| | Acceptance and compliance | Reference |
|---|---|---|
| Propagation delay | x | x |
| Skew | x | x |
| DC loop resistance | | x |
| NEXT | x | x |
| ELFEXT | x | x |
| Attenuation | x | x |
| Return loss | x | x |
| Longitudinal conversion loss | | x |
| Transfer impedance | | x |
| Coupling attenuation | | x |
| Shield dc resistance | | x |
| Conductor map | x | x |
| Length | x | x |
| Continuity of conductors and shields | x | x |

Adapted from draft ISO/IEC 11801:2000, March 1999, Table A.1.

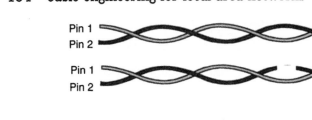

**Fig. 10.2** Wire map problems.

The test report should have a unique identity number for the cable circuit, the project name or address, the time and date of the test and the name of the operator. The file should be saved electronically in the same file format that is supported by the cable management software supplied by the tester manufacturer.

As to the test parameters themselves; they are mostly self-explanatory. All of the terms are explained in chapter 9. Wire Map is the check to ensure that every wire is connected through to the appropriate connector pin on the far end of the circuit. It looks for short and open circuits, crossed pairs, split pairs and tip-ring reversals. Figure 10.2 shows the possible wire map problems.

## 10.4   Testing standards

In the beginning, the only published test standard was TSB 67, i.e. *Telecommunications systems bulletin* no. 67 of ANSI/TIA/EIA 568A. In America TIA/EIA 568A TSB 95 and TIA/EIA 568A Addendum no. 5 have added to this standard. Both expect higher performance from the cable system and are referred to as enhanced category 5. They add parameters such as ELFEXT and delay skew to be measured. In 2000, TIA/EIA 568B:2000 will be published which will bring all previously published TSBs and amendments together. Table 10.1 details the exact requirements.

A great deal of work has been done by the IEC to catch up with the international standards for testing cable systems and components. Figure 10.3 shows how the families of test standards interrelate.

ISO 11801 2nd edition:2001 refers to IEC 61935 part 1 for copper cable system testing. It also refers to IEC 61280 parts 1, 2 and 3 and IEC 60793 part 1 for optical testing. ISO 14763, which relates to the planning and installation of cable systems abandoned its own part 4, *copper cable testing*, and refers now instead to IEC 61935. Within CENELEC, EN50174 is the cabling system design and test standard. For testing details it will refer to the as yet un-named EN 50xxx, which in turn will look to IEC 61935 in an effort to seek international compatibility between the standards.

The full names of the standards are:

• IEC 61935 generic specification for testing of generic cabling in accordance with ISO/IEC 11801 — part 1: test methods.

The following are component tests:

• IEC 61935 generic specification for testing of generic cabling in accordance with ISO/IEC 11801 — part 2: patch cord and work area cord.
• IEC 61156-1 multicore and symmetrical pair/quad cables for digital communications — part 1: generic specification.
• IEC 60603-7 Amendment 1. Detail specification for connectors 8 way including fixed and free connectors with common mating features. Amendment 1: test methods and related requirements for use at frequencies up to 100 MHz.

## 10.4.1   Accuracy

There is a great temptation, when presented with a numerical readout of a parameter on a digital readout, to accept its validity at face value. If it says 3 dB it must mean 3 dB. Hand-held testers are not laboratory instruments maintained in a controlled environment and

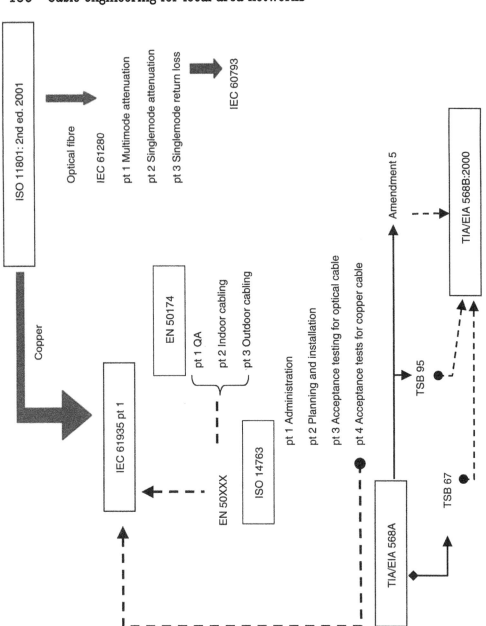

**Fig. 10.3** Family of test standards.

calibrated regularly. Every testing device has a working tolerance, and installers and end-users alike must understand this. Every measured value has a tolerance, and some parameters have the compound errors of several measurements. For example, an ACR reading will include the combined errors of the Attenuation and the NEXT reading. The length reading will have the compound errors of the time domain reading itself plus the uncertainty factor of the value of NVP used.

The standards have tried to bring a level playing field to the testing issue by including levels of measurement error allowed. For category 5 testing we had levels I and II. Cat 5e requires level IIe and category 6 requires level III. There will also be a level IV. Table 10.3 gives the tolerances allowed for different levels.

A potential problem which has arisen with the advent of 250 MHz cable systems, and their subsequent testing, is the manufacturer's need to match plug and socket uniquely from their own product range. The spirit of the standards is the intermateability of connectors from different manufacturers giving the same category of performance when like connectors are used. This does work for connector pairs up to and including cat 5e. In other words a cat 5e

| Table 10.3 Accuracy levels for testers (adapted from TSB 67) | | |
|---|---|---|
| | Channel | Basic link |
| Accuracy level I | | |
| Attenuation | ±1.3 dB | ±1.3 dB |
| NEXT | ±3.4 dB | ±3.8 dB |
| Length | ±1 m ±4% | ±1 m ±4% |
| Accuracy level II | | |
| Attenuation | ±1.0 dB | ±1.0 dB |
| NEXT | ±1.5 dB | ±1.6 dB |
| Length | ±1 m ±4% | ±1 m ±4% |
| Accuracy level IIe* | | |
| NEXT | 2.8 dB | 1.6 dB |
| Accuracy level III* | | |
| NEXT | 2.6 dB | 1.7 dB |
| *Proposed value. | | |

plug from one manufacturer, when mated to a cat 5e socket from another manufacturer, will still give a combined electrical performance conforming to cat 5e requirements. Unfortunately it appears that the 250 MHz requirements of cat 6 means that all plugs and sockets have to be produced as matched pairs. On one level it means that a category 6 cable system could only come from one manufacturer, no mix-and-match solutions are likely to work. On the other hand it calls into question the accuracy of any generic interface connector supplied with a hand-held tester when connected to a category 6 cable system. Tester manufacturers have had to respond with connector manufacturer specific interfaces for every connector encountered. This is a far from ideal solution but seems inevitable until a truly generic cat 6 connector or test interface can be devised.

## 10.4.2   Channels and links

The final consideration when testing copper cabling systems is, 'what is being tested?' For field testing there are two valid scenarios; channel testing and basic link testing.

Figure 10.4 demonstrates the alternative configurations within a cabling link. We can see that the *'channel'* has a specific meaning.

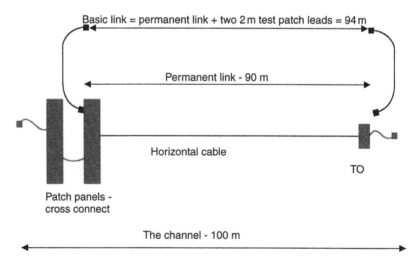

**Fig. 10.4** Channel and links.

It is everything from the end of one work area cable to the end of the other one at the far end. Under cat 5, cat 5e and cat 6 (class D and E) rules, this cannot exceed a total distance of 100 metres. Up to ten metres is allowed for patch cables and so 90 m are allowed for the fixed cabling. The 'permanent link' is as it says, everything that is permanent in the installation and is normally from the patch panel, through the horizontal cabling to the telecommunications outlet. It would usually include the transition or consolidation points as they are considered permanent or semi-permanent. Note that TIA/EIA 568B infers that the test is done from either the TO or the transition point but the latter is unusual in practice.

The permanent link is really a design exercise. To measure it a connection has to be made to the test instrument. The permanent link, connected at each end by two metres of test patch cord to the hand-held tester is termed the *'basic link'*. It is the basic link that is normally measured in most installations. Note that the 2 m patch leads come with the tester, they are not the normal patchleads or work area cables. Testing the basic link is popular because it is simple and repeatable; it is after all the permanent part of the cable system that is being measured.

Some users will require that the complete channel be tested, as this is what will be used in real life. There are disadvantages to this approach however.

- All the patch cords that can potentially be used in a cabling system are never all delivered on day one, so what is to be tested? Generally a five metre standard patch cord will be used on each end of the link to emulate a worst case channel but this no more reflects reality than testing a basic link.
- At some point in the future a patch cord might be changed or a jumper moved in the cross connect. The resulting cable channel is then an unknown and untested quantity.
- The hand-held tester normally presents an RJ45 plug. For a channel, also ending in an RJ45 plug, a channel adapter must be used.

To allow for these two approaches to testing, the standards all give tables of specifications for both channel and basic/permanent link performance. The channel performance is always slightly lower than the basic link performance because of the allowance for longer patch-leads and the possibility of two cross-connects being included. When using a hand-held tester there is the option to instruct it to test to channel or basic link parameters.

# 11

# Optical cable technology — optical fibre

## 11.1 Introduction

Optical fibres can be defined in several ways, for example multimode and singlemode, graded index and step index etc. All optical fibres have the following in common; a core of a transmissive medium with a certain refractive index and a cladding of a material with a lower refractive index. It is this difference in refractive index that constrains the light to mostly remain within the fibre and not escape out of the sides. The core can be made of glass or plastic and the cladding can also be made of glass or plastic. The majority of optical fibres are all silica, i.e. glass, but there are fibres with a glass core and a plastic cladding or even composed entirely of plastic. The vast majority of optical fibres used in communications are made of all-silica; plastic fibre has often been portrayed as the next generation low-cost alternative, but as yet it has not found many applications beyond the speciality market.

Experiments on optical fibre in the 1960s demonstrated that it was feasible to manufacture a glass waveguide and send optical signals down it. It was not until the late 1970s that the mass production of low loss fibre became a reality. Optical fibres are firmly established as the principal transmission medium for all long-haul, cable based telecommunications networks. Optical fibre has been used in datacommunications, LANs and premises cabling systems since the late 1980s and is totally accepted as the backbone cabling of choice. Fibre to the desk, for the time being, is still restricted to those who need the additional performance factors of optical fibre, namely long distance, total security and freedom from EMI/EMC.

# 11.2   Why use optical fibre?

Optical fibre has been proposed as an alternative to copper cable, so an initial question must be; when would you use it? The principal advantages of optical fibre over copper are: —

- *It has a much lower attenuation.* 100 m of category 5 cable would have an attenuation of around 20 dB. 100 m of optical fibre has an attenuation of not more than 0.3 dB, i.e. approximately 85–100 times less. The corollary of this is that signals transmitted on optical cables can go much further than on copper cables before they have to be regenerated. When telecommunications systems operated on coax, repeater spacings of 500 m were the norm. Optical cables have repeater spacings at 25–40 km typically, i.e. a 50–80 times improvement in cost, reliability and accumulated noise and jitter problems.
- *Optical fibre has a much higher bandwidth.* The useable bandwidth of 100 m of category 5 is 100 or 200 MHz for category 6. 100 m of optical cable has a minimum bandwidth of 1600 MHz. Singlemode fibre would have a bandwidth measured in hundreds of gigahertz over such a distance. With a larger bandwidth the information carrying potential of optical fibre is much greater.
- *Freedom from electrical interference.* Optical fibre does not suffer from any EMC/EMI related problems. It is immune to electrical interference and it does not radiate any interference. Lightning, electrostatic discharge and electrically noisy machinery have no effect on optical fibre.
- *Security.* As fibre does not radiate any energy outside of the cable it is impossible to 'electronically eavesdrop' onto any of the data flowing down the cable. This makes optical fibre extremely

popular with military establishments. It is possible to manipulate a fibre to the extent of extracting some of the light within it, assuming one has unfettered access to the bare fibre and precision engineering skills, but fibre can be considered totally secure for all practical purposes. In some military establishments, for the most secure circuits, the cables are laid in open or see-through trunking that in itself is laid in corridors or other places where it is permanently open to view.

- *Freedom from earth loops and fault currents.* The voltage appearing on the grounding system in one building may differ from that of another building. Even if this is only a fraction of a volt it will cause troublesome earth loop currents to circulate through copper cables. Non-metallic optical cables will not allow or cause earth loop currents to flow through them. Non-metallic optical cables are also totally immune from lightning strikes and will not conduct surge currents into a building. Any metallic elements within an optical cable, such as armouring or strength members would still have to be grounded at the entrance to a building so they don't become a route for fault currents.

- *Optical cables are small and light.* A duplex, indoor grade, optical cable is about the same size as a category 5 copper cable, i.e. around 5 mm in diameter, but the optical cable will be much lighter and stronger. A backbone 24-fibre cable will be many more times smaller and lighter than say a 300-pair copper cable and will have thousands of times the information carrying capacity. The smaller size and weight of optical cables may be important in older buildings where there is simply insufficient space for large volumes of heavy copper cables. The low weight and size of optical cables promises many opportunities for the use of optical fibre in the aerospace, military and maritime markets.

With optical fibre having so many virtues why do we use copper cables at all? The answer comes down to cost. For long-haul routes, i.e. anything longer than 2 km, there is no question that optical fibre is the most cost effective method of communications. For campus and backbone cabling, which covers distances from 100–3000 m, optical fibre is usually the cheapest cabling option unless the communications requirements are very trivial or purely telephony based. For the horizontal cabling, i.e. the 'to-the-desk' cable, copper cable is still the most cost-effective route, especially with cat5e and category 6 cabling promising gigabit transmission speeds over these distances.

It doesn't just come down to the cost of cabling. Duplex multimode fibre to the desk costs about 50% more to supply and install than UTP and about 30% more than screened cable, and the optical transmission equipment is still in the range of being three to four times the cost of its copper equivalent, even for doing the same job. The cost-models of fibre to the desk can look more attractive for the larger site when home run or centralised optical cabling is used. Centralised cabling (see TSB 72 Standard) allows 300 m runs of optical cable from the user to the centralised equipment room, bypassing the need for horizontal cross-connects or floor distributors.

Although there are many commercial examples of fibre to the desk installations, at present, the vast majority of premises cabling systems utilise category 5 copper cable to the desk and optical fibre in the backbone and campus wiring. Those that do specify fibre all the way to the desk are the class of user who need the additional benefits of optical fibre, with security usually being at the top of the list, but freedom from EMI and fewer length restrictions are the second and third most quoted reasons.

# 11.3   Fibre manufacture

There are three principle methods for the production of fibre. A thorough understanding of optical fibre manufacturing methods is not necessary to be able to use fibre and so the three methods are only summarised here.

- OVD   Outside vapour deposition. A glass target rod is placed in a device like a lathe. As it rotates a vapour of germanium and silicon tetrachloride is sprayed onto it. As the gas nozzle passes up and down the rod a layer of glass-like material builds up. Varying the composition of the vapour, particularly with regards to germanium, can vary the eventual refractive index profile of the fibre at this stage. Eventually the preform is finished when it reaches about one metre long by about 15 cm diameter. The target rod is withdrawn and the preform is sintered in an oven to consolidate the glass and drive off impurities such as water vapour. The preform is placed vertically in a drawing tower. The end is heated and pulled downwards by a tractor feed. By use of a sophisticated feedback control system the finished optical fibre is drawn off the preform. Before the fibre is wound onto a reel however an ultra-violet cured acrylate material is applied to give the fibre what is known as its primary coating. The fibre is drawn off with a diameter of 125 μm. The acrylate coating brings the diameter up to 250 μm. The primary coating is essential to give the fibre enough mechanical protection to survive handling, testing and eventually cabling.

- VAD    Vapour axial deposition. In the VAD method the preform is built up by the chemical vapour being sprayed onto the end. Thus the preform builds up axially rather than concentrically.

- IVD    Inside vapour deposition. This process is also known as MCVD (modified chemical vapour deposition) and PCVD (plasma chemical vapour deposition). In this process the starting point is a glass tube and the chemical vapour flows through the middle of it, building up concentric layers from the inside. At the end of the process the preform is collapsed down before sintering and drawing.

The world market share of the three processes is similar, though PCVD is the smallest.

# 11.4   Multimode and singlemode, step index and graded index

When the rays of light enter the core of an optical fibre, as seen in Fig. 11.1, there are numerous routes they can follow. Some can go straight down the middle and some will be reflected from side to side.

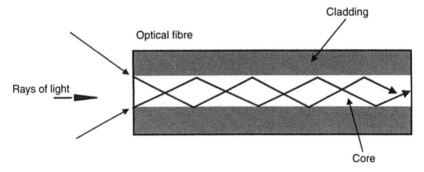

**Fig. 11.1** Optical fibre.

Table 11.1 Theoretical number of modes supported in different optical fibres

| Fibre type | 850 nm operation | 1300 nm operation |
|---|---|---|
| Singlemode | 0 | 1 |
| 50/125 multimode | 350 | 150 |
| 62.5/125 multimode | 1000 | 450 |

Each of these rays of lights can be considered as modes, and a large-core fibre can support hundreds of these modes, hence the term, multimode. Table 11.1 gives the theoretical number of modes supported by different kinds of optical fibres.

These modes will all travel different distances to reach the far end of the fibre, and as they travel at the same speed they will all arrive at different times. Light travelling down the centre of the core will arrive at the far end before the rays of light that have bounced from side to side within the core. This is the first problem that optical fibres have to overcome, modal dispersion. If we imagine a series of sharp pulses of light entering the core at the transmitter end, representing a stream of digital 'ones' and 'zeroes', then the effects of modal dispersion will cause the sharp edges of the pulses to spread out to the point where they will merge and the original data is lost. This is the bandwidth limiting effect of modal dispersion.

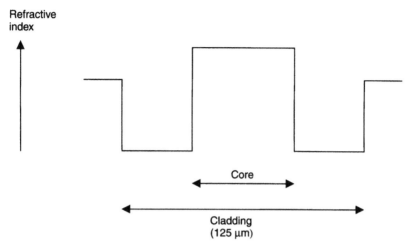

Refractive
index

Core

Cladding
(125 µm)

**Fig. 11.2** Step index optical fibre.

This simple kind of fibre is called a step index fibre. The profile through the core and cladding, shown in Fig. 11.2, looks like a step as the refractive index of the cladding changes sharply to that of the core, where it remains the same until it drops down again at the cladding at the other side of the core.

The first optical fibres were step index, and today singlemode and some of the larger core specialist fibres are still step index, but in communications the next major development was the introduction of graded index fibre.

Graded index fibres, seen in Fig. 11.3, take advantage of the fact that the refractive index controls the speed at which the light travels in the fibre. The refractive index profile of the core can thus be engineered so that at the centre the refractive index is highest. So at the centre, where the rays of light have the least distance to travel, they are constrained to travel at the slowest speed. The modes travelling nearer to the periphery of the core are in a region of lower refractive index. Thus the outer modes which have to travel further can progress at the higher speed. The total effect of the graded index core is that the various modes of light are more likely to arrive at the far end of the fibre grouped much closer together in time. The digital data stream will therefore suffer much less modal dispersion and so

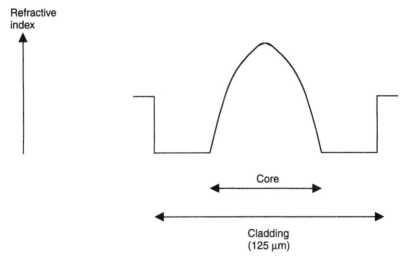

**Fig. 11.3** Graded index optical fibre.

the bandwidth of graded index fibre is much higher than that of a multimode step-index optical fibre.

Singlemode fibre, Fig. 11.4, sometimes called monomode fibre, works in yet a different way. Singlemode fibre is a step index fibre but the diameter of the core is much smaller than in a multimode fibre. A multimode fibre has a core diameter of typically 50 or 62.5 µm, or millionths of a metre, but a singlemode fibre has a core diameter in the range of only eight to ten microns. This has the optical effect that for light rays above a certain wavelength only one mode will propagate down the core, hence the expression, singlemode. Singlemode fibre has a much larger bandwidth than multimode fibre because it does not suffer from modal dispersion. The wavelength at which singlemode operation comes into effect is called the cut-off wavelength, and is typically around 1250 nm. The fibre cut-off wavelength can be between 1190 and 1330 nm for typical singlemode fibre but it is dependent upon length and bends imposed on the fibre so the working value is often referred to as the cable cutoff wavelength.

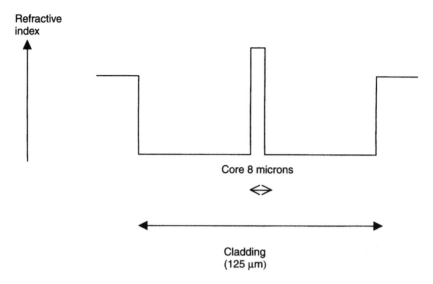

Fig. 11.4 Singlemode optical fibre.

# 11.5  Optical fibres for datacommunications and telecommunications

Although there are many kinds of optical fibre available, three in particular represent 99.99999% of the communications industry. First of all there are two multimode fibres, 50/125 and 62.5/125. The first figure represents the diameter of the core in microns, and the second figure gives the overall diameter of the cladding.

The third type is singlemode fibre, which might also be described as 9/125. Figure 11.5 demonstrates the relative sizes of the fibres.

But why do we need different kinds of fibres? As a general rule of thumb the smaller the core, then the higher the bandwidth (hence more data carrying capacity), the lower the attenuation (hence longer transmission distances) but, the harder it is to launch light into and the tighter the tolerance required on splicing and connectorising.

Multimode fibres can transport a data rate of at least 155 Mb/s over 2 kms using a cheap LED (light emitting diode) as a source. Until now this has been more than adequate for LANs working on premises

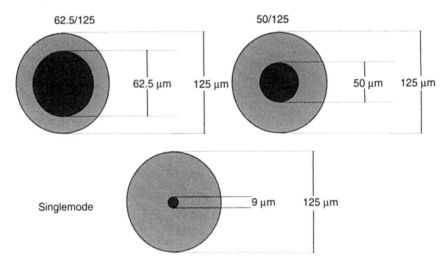

**Fig. 11.5** Different kinds of fibres used in communications.

cabling systems. Singlemode fibre can transmit gigabit/second rates of data over large distances, typically 40 km between repeaters, but it requires a laser to launch the light into the small core.

Traditionally lasers have been very expensive. They are intended for the long haul telecommunications market where high output power, stability and reliability are the expected norm. A telecommunications grade laser can cost anything from $2000 to $4000. It can modulate data at gigabit speeds and transmit data over 40 km or more. The cost of the laser is not really an issue in the telecommunications business. The cost is very small relative to the entire infrastructure costs of a route measured in tens or hundreds of kilometres. This kind of cost and performance is way over the top for Local Area Networks where a 2 km route would be considered long — the average is more in the region of 300 m, and until recently had a speed requirement no greater than the 100 Mb/s of FDDI and Fast Ethernet. LEDs cost in the order of tens of dollars for 850 nm operation and hundreds of dollars for 1300 nm, and can transmit up to 155 Mb/s over 2 km of multimode optical fibre.

The cost of the transmission equipment is the reason why we have the traditional split of multimode fibre up to 2 km and singlemode fibre for distances beyond that. In the telecommunications business all optical fibre is now singlemode. As most premises cabling systems are less than 4 km in diameter, the optical fibre used has been almost exclusively multimode. In terms of cost of cabling, singlemode fibre is actually the cheapest, followed by 50/125 and then 62.5/125. But the overall systems costs are driven by the transmission equipment which traditionally prices singlemode equipment with the telecoms market in mind.

A number of issues have recently arisen which challenge this paradigm.

- A need for faster transmission speeds, e.g. gigabit Ethernet, ATM, high speed Token Ring and fibre channel. LEDs cannot modulate much beyond 200 MHz, and so cannot cope with gigabit transmission speeds.
- The advent of a cheaper-to-produce family of lasers. These are known as VCSELs (vertical cavity surface emitting lasers) and they can be produced for tens of dollars. They do not have the speed, power or tight spectrum of a telecommunications laser such as a DFB (distributed feedback laser) but what they can do is more than enough to meet the gigabit speeds over the few kilometre requirements of LANs.

VCSELs have so far only been produced to work at 850 nm over multimode fibre, but 1300 nm versions for multimode and singlemode fibre are being developed. They promise to revolutionise the use of singlemode fibre in LANs.

# 11.6    Wavelengths of operation

Optical fibre has a different performance depending upon the wavelength of light used. Figure 11.6 shows how the attenuation, in dB/km, changes according to the wavelength of light travelling down the optical fibre.

Some of the differences are caused by impurities in the glass absorbing light at particular wavelengths. The scale starts at 800 nms which is the start of the infra red band. It is just at the limits of

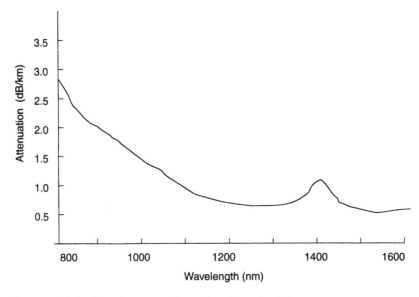

**Fig. 11.6** Wavelength versus attenuation for optical fibre.

detection for the human eye. If we moved to the left, to a shorter wavelength we would get to visible red light at about 600 nm. The sensitivity of the human eye peaks around 500 nm, which matches the yellow light of the sun. Beyond that we move into blue light and then onto UV.

Optical fibre operation covers the band from 800 to 1600 nm. We can see that there are several minima; one at 1300 nm and the other at 1550 nm. 850 nm is not particularly a minimum but it is a convenient window to use because of the availability of LEDs operating at this wavelength. The three wavelengths are called windows of operation:

- First window — 850 nm.
- Second window — 1300 nm.
- Third window — 1550 nm.

Again to generalise; the longer the wavelength the lower the attenuation and the higher the available bandwidth but also the higher the cost of the laser. 1550 nm operation gives the option of very long distance transmission in the order of hundreds of kilometres before signal regeneration is required. 1550 nm operation is used for very long transcontinental and transoceanic routes but the 1550 nm lasers are very expensive.

## 11.7    Bandwidth and dispersion of optical fibre

The bandwidth of optical fibre is a measure of its information carrying capacity. The unit is MHz.km. The longer the distance then the less bandwidth is available.

The limiting factor for bandwidth in optical fibre is dispersion, which essentially means that the components of the optical signal arrive at different times. As the speed of transmission goes up so the pulses of light launched into the fibre have to be squeezed closer and closer together in the time domain. Eventually dispersion causes the pulses of light to merge into each other and the original transmitted information is lost. Optical fibre exhibits different types of dispersion. Multimode fibre suffers from modal, chromatic and differential modal dispersion. Singlemode fibre suffers chromatic and polarisation mode dispersion plus more esoteric problems that have come to light with the advent of dense wavelength division multiplexing (DWDM). These are self-phase modulation, cross-phase modulation, modulation instability and four-wave mixing.

### 11.7.1    Modal dispersion

As already discussed, multimode fibre supports hundreds of different modes or rays of light. As each one can take a different path through the core they will arrive at different times at the far end, hence causing dispersion. Graded index multimode fibre overcomes this problem to some extent. Singlemode fibre will not suffer this problem, as long as the operational wavelength is above the cut-off wavelength, as only one mode is propagated.

### 11.7.2    Chromatic dispersion

Also known as material or intramodal dispersion. The refractive index of glass is dependent upon the wavelength of light measured. To turn this around, every wavelength of light travels at a slightly different

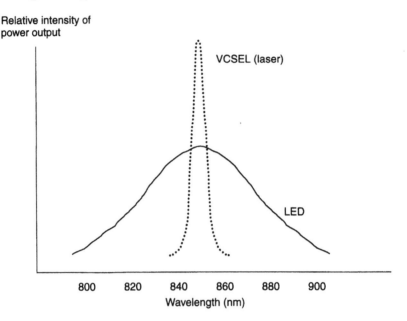

**Fig. 11.7** Spectrum of LEDs and lasers.

speed within the fibre core. The light producing device, be it LED or laser is not monochromatic, i.e. the energy it transmits is spread over a number of wavelengths, even though the centre peak may be at 850 or 1300 nm. The spectrum of an LED is much broader than a laser. Figure 11.7 shows the spread of energy from different types of transmitters.

When the bandwidth of an optical fibre is quoted it is based on the modal dispersion, as this can be measured quite accurately. Chromatic dispersion depends upon the spectrum of the transmitting device. If the source were totally monochromatic then there would be no chromatic dispersion. This is impossible in reality and we can assume from the spectrum of an LED that chromatic dispersive effects could be severe. Optical fibre specification sheets will give a value for chromatic dispersion in picoseconds/nm.km for optical fibre in the form of an equation:

$$dispersion = D(\lambda) \approx \frac{S_0}{4}\left[\lambda - \frac{\lambda_0^4}{\lambda^3}\right]$$

[11.1]

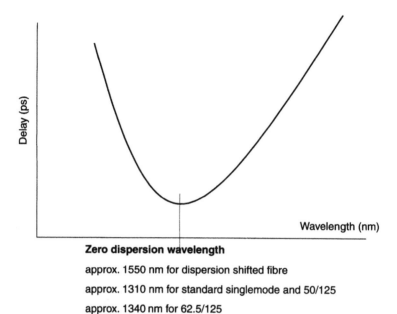

**Zero dispersion wavelength**

approx. 1550 nm for dispersion shifted fibre

approx. 1310 nm for standard singlemode and 50/125

approx. 1340 nm for 62.5/125

**Fig. 11.8** Fibre chromatic dispersion.

Equation 11.1 has the units of picoseconds per nanometre.kilometre. So the wider the spectrum of the source, in picoseconds, and the longer the cable, in kilometres, then the more dispersion, in picoseconds, there will be.

In the equation $\lambda_0$ is the zero dispersion wavelength and $S_0$ is the zero dispersion slope. The slope of the dispersion is not monotonic, that is it changes direction. Figure 11.8 makes this easier to understand. At a certain wavelength the delay is at a minimum. This is the zero dispersion wavelength and the slope at that point is the zero dispersion slope. For a typical 62.5/125 fibre, equation 11.1 is valid in the range 750–1450 nm and $\lambda_0$ is between 1332 and 1354 nm with a zero dispersion slope of less than 0.097 ps/(nm$^2$.km).

## Differential mode delay and laser launch bandwidth

Bandwidth of multimode optical fibre has traditionally been measured by inserting light from an LED or a laser modified to act like an LED, and increasing the frequency of modulation until the output power at the far end of the fibre falls by 3 dB, although the result is always normalised back to one kilometre, hence MHz.km. The LED is like a searchlight compared to the diameter of the core and so the core, and some of the cladding is completely filled with light. This condition is referred to as overfilled launch (OFL). Lasers were not much used with multimode fibre due to their high cost, but with the advent of VCSEL technology this is changing, with the main instrument of change being the gigabit Ethernet standard 1000baseSX, ratified in the summer of 1998, which uses 850 nm VCSELs.

This has led to a renewed interest in the bandwidth capabilities of optical fibre when operated with lasers. Many people presumed that the bandwidth of the fibre would be the same, regardless of the method of launching light into them, but this has not proven to be the case. The main difference with lasers is that a very small spot size of light is launched into the core, and whatever part of the core is being used or illuminated by that spot of light has an effect on the bandwidth performance. LEDs flood the core and modal effects and differences are averaged out over the length of the fibre.

There are reasons why OFL and laser launch bandwidth should be the same, but only when using a perfectly-made fibre. Perfect in this sense means a flawless refractive index profile. In many instances the refractive index has a dip right in the middle, which is an artefact of the manufacturing process. The bandwidth performance of laser launched fibre can therefore be unpredictable. Figure 11.9 demonstrates the non-correlation of a group of multimode optical fibres that have had their bandwidth measured by both the LED overfilled launch method and by a laser-launched beam of light. The non-correlation represented by the graph demonstrates how bandwidth measured by one technique has little connection to the bandwidth available to the other technique.

The variations away from the ideal refractive index profile mean that a spot of light will be delayed more or less than desired as that spot is scanned across the core. The effect of this is known as differential mode delay (DMD). DMD reduces the bandwidth available and for bandwidth limited protocols such as gigabit Ethernet (most other protocols are attenuation limited) it leads to reduced length allowed in the cabling. Figures 11.10 and 11.11 show the effect of scanning a

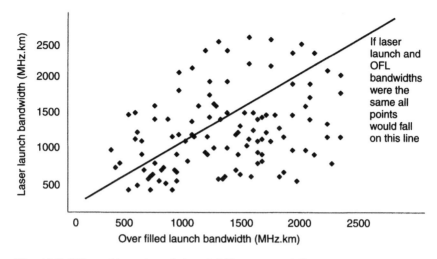

**Fig. 11.9** OFL and laser launch bandwidth non-correlation.

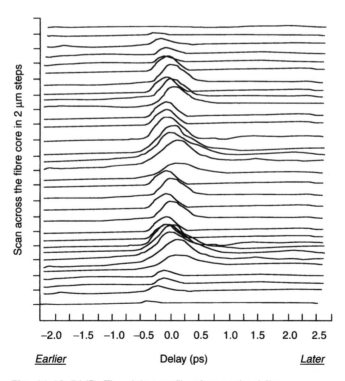

**Fig. 11.10** DMD. The delay profile of an optimal fibre.

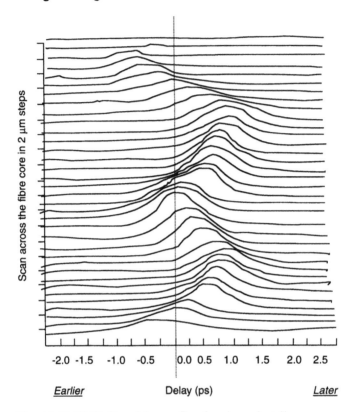

*Earlier*          Delay (ps)          *Later*

**Fig. 11.11** DMD. The delay profile of an imperfect fibre.

2 μm wide beam across the core of a perfect and an imperfect multimode fibre. Figure 11.12 shows the refractive index profile of an imperfect, but common, multimode optical fibre.

The 1000baseLX gigabit Ethernet standard requires an offset launch lead, which ensures that the laser spot is never launched directly into the centre of the core, which is what an optical connector would normally try to achieve. For 62.5/125 fibre the offset is between 17 and 23 μm from the centre of the core; for 50/125 fibre the offset is between 10 and 16 μm. Note from Figure 11.13 that the launch lead from the laser transmitter is singlemode fibre and the offset is effected through an offset device joining the singlemode to the multimode exit fibre.

A new generation of optical fibres has been designed to overcome this problem; i.e. a perfect index profile has been sought. An example

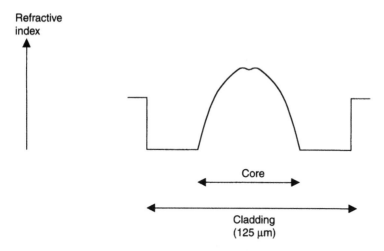

**Fig. 11.12** Imperfect, but common, refractive index profile of a multimode fibre.

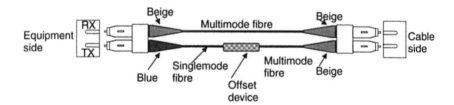

**Fig. 11.13** Offset-launch mode-conditioning patch cord assembly for 1000baseLX.

of this new kind of fibre is the Corning Infinicor CL, which is optimised for laser launch conditions.

# 11.8 Attenuation

Attenuation in optical fibre is measured in decibels or decibels per unit length, usually kilometres, e.g. dB/km. Optical fibre glass is extremely pure. The 3 dB power point, i.e. the point at which half of the power has been absorbed is at least 1 km for multimode optical fibre, for singlemode fibre it is nearer 10 km or more. In comparison, window glass absorbs half of the light passing through when only three centimetres thick!

Optical fibre attenuation can be attributed to two basic loss mechanisms:

- Intrinsic.
- Extrinsic.

Intrinsic loss is a property of the glass itself. There is Rayleigh scattering that is caused by microscopic irregularities in the glass reflecting light back towards the source. There is also absorption attenuation, which is mostly due to impurities in the glass, such as residual water ions, which absorbs light more strongly at particular wavelengths.

Extrinsic attenuation is caused by external forces acting upon the fibre and can be grouped into macrobending and microbending. Macrobending is due to large scale visible bending of the fibre when the minimum bend radius is infringed. Under this condition, light, of a low enough wavelength, e.g. red, can actually be seen leaking out of the side of the fibre. Macrobending is dependent upon wavelength, with longer wavelengths suffering more from this phenomenon. Macrobending in multimode fibre represents the higher order modes being lost. Although this represents attenuation, as light power is lost, it also increases the bandwidth, as the highest order modes are the largest contributors to modal dispersion.

Microbending is the effect of very small scale distortions on the surface of the fibre caused by the local effects of primary coating, tight buffering or cabling. Microbending is not particularly dependent upon wavelength for multimode fibre and is slightly dependent upon wavelength for singlemode.

## 11.9   Singlemode fibre

Singlemode fibre used to be the preserve of telecommunications but its presence is now being felt in the world of LANs, especially in the larger campus where cable runs may be from 500 to 3000 m. Laser optimised multimode fibre is all very well but standard singlemode fibre is still the cheapest and most capable type of fibre available.

As already discussed, singlemode fibre has a very small core, typically around eight to nine microns. The small size of the core constrains light above a certain wavelength to only propagate in one mode. Singlemode fibre is still referred to sometimes as monomode. In data sheets however it is not the core size which is given but something called mode field diameter. All of the light does not travel entirely within the core, some of it travels in the cladding right next to the core. If one observed the light intensity across a singlemode fibre a cone of light, slightly larger than the core, would be seen. The diameter of this cone of light is called the mode field diameter, and it has a typical value of between 9.5 and 10.5 µm.

## 11.9.1   Dispersion effects in singlemode fibre

Singlemode fibre, whilst escaping modal dispersion by its very nature, suffers from a few problems of its own. Singlemode and multimode share the problems of chromatic dispersion and waveguide dispersion. The latter is a differential effect of light travelling partly in the core and partly in the cladding. Engineering the refractive index profile at the core-cladding boundary can change the zero dispersion point. Singlemode fibre also suffers from polarisation mode dispersion (PMD).

## Polarisation mode dispersion

Light, like all other electromagnetic energy, is composed of two orthogonal waveforms that normally travel together. If something in the transmission system affects this so that the two modes change with respect to each other then we call the light 'polarised'. If we look through polarising sunglasses, at the sky or through a car windscreen, we are only seeing one of the polarisation modes. If we turn the glasses through ninety degrees then we will see something quite different. Some 3-D cinemas work by transmitting two pictures on the screen, each polarised ninety degrees with respect to the other. Viewed normally we would see two pictures superimposed that created an unwatchable effect. But when viewed through polarising glasses with each lens polarised at 90° to the other then one signal is blocked out from each eye and a three dimensional illusion is perceived.

If the core of an optical fibre were perfectly round then the two orthogonal EM fields would be unaffected. But no fibre can achieve totally perfect circularity and so some polarisation effects will occur to the extent that one mode will arrive slightly later than the other. The effect is also exacerbated by induced stresses in the fibre from cabling and installation. PMD can adversely affect long analogue video systems and high-speed digital systems.

The answer to PMD is accurately made fibre and low-stress cable designs.

## 11.9.2    Types of singlemode fibre

Standard singlemode fibre (SMF) is optimised for use at 1310–1312 nm, principally meaning that the chromatic zero dispersion point is placed around this wavelength. SMF is still the principal fibre in use today, but the advent of optical amplifiers and a desire to go longer distances stimulated users to look again at 1550 nm performance. 1550 nm operation has a lower attenuation, typically 0.2 dB compared to 0.34 at 1300 nm, but the dispersion is much worse, e.g. 18 ps/nm.km.

Optical amplifiers, known as EDFAs (erbium doped fibre amplifiers) work at 1550 nm. In the late 1980s dispersion shifted fibre (DSF) was introduced. DSF has the zero dispersion point centred at 1550 nm.

Increasing optical output powers revealed a problem with singlemode fibre, especially DSF, based on a non-linear performance. At high power, the fibre exhibits a non-linear response, and in any system, electronic or optical, a non-linear response leads to a number of intermodulation effects. For DSF optical fibres these can be summarised as self-phase modulation, cross-phase modulation, modulation instability and four-wave mixing. Of these, four-wave mixing is regarded as the most troublesome. Four-wave mixing generates a number of 'ghost channels' which not only drain the main optical signal but also interfere with other optical channels. For wavelength division multiplexing (WDM) this is a severe problem, for example a

32 channel WDM has a potential 15872 mixing components. Four-wave mixing is most efficient at the zero dispersion point.

A new fibre was therefore introduced called non zero dispersion shifted fibre (NZDSF). NZDSF has the zero dispersion point moved up to around 1566 nm. This is just outside the gain band of an EDFA. So a small amount of dispersion is introduced into the system but at a controlled amount and with minimised four-wave mixing.

For the vast installed base of SMF, it is possible to add quantities of dispersion compensating fibre (DCF) to equalise the dispersion effects. DCF has an opposite dispersive effect to SMF and can reverse pulse spreading to some extent.

Another approach is to make the core area of the NZDSF larger. A larger core will have a lower energy density and so four-wave mixing effects will be much lower. So-called large effective area fibres have a core area around 30% larger than conventional singlemode fibres and this gives a much lower energy density. There has to be a limit of course. Making the core larger and larger would eventually return it to a multimode fibre, but the principal limiting factor is the larger dispersion slope that a larger core area brings. A larger dispersion slope means that the amount of dispersion will be very different according to which wavelength is being used.

Designing long-haul singlemode optical systems is now a complex economic exercise that has to balance competing technologies with an acceptance of the reality of the existing installed base of SMF.

For LANs and premises cabling systems, conventional 1310 nm singlemode fibre will remain the most viable and economic single-mode solution for some time yet.

## 11.10    Plastic optical fibre

Plastic optical fibre has long held the promise of a low cost, easy to install communications medium that offers all the benefits of optical fibre with the ease of termination of copper. Unfortunately plastic fibre is not yet proven to be cost competitive or to exhibit sufficiently high bandwidth or low enough attenuation to make it a serious rival to either glass fibre or copper cable. Developments continue however and it would not be wise to write off plastic fibre just yet.

Plastic fibre available today is step index, which by its very nature limits the bandwidth available. Current designs are based on a material called PMMA (polymethyl methacrylate). Step index plastic optical fibre (SIPOF) today has a best bandwidth of 12.5 MHz.km and an attenuation of 180 dB/km. Compare this to the 500 MHz.km bandwidth and 1 dB/km attenuation available from 50/125 glass optical fibre.

The manufacturing costs of PMMA fibre are thought to be about the same as for conventional glass optical fibre, but SIPOF currently sells at a premium compared to glass or all-silica fibre. The thermal stability of PMMA is also questionable. High temperatures combined with high humidity can raise the attenuation of the fibre significantly.

SIPOF fibres are available in sizes of 500, 750 and 1000 μm total diameter. Most of this is a PMMA core with a thin layer of fluorinated PMMA for the cladding.

PMMA has attenuation minima occurring at 570 nm and 650 nm. The theoretical minimum attenuation achievable at these wavelengths is 35 and 106 dB/km respectively. However there are no 570 nm sources available so it is only practical to use the 650 nm window.

Deuterated PMMA has been proposed as an advancement. It can reduce attenuation down to 20 dB/km in theory but this has not been achieved in practice. Deuterated PMMA is also very expensive to produce.

To really improve plastic fibre a graded index version has to be produced to overcome the poor bandwidth properties of SI-POF. Graded index plastic optical fibre, or GI-POF, offers the potential of 3 Gb/s transmission over 100 m and 16 dB/km attenuation at 650 nm. Even 1300 nm operation may be possible with next generation materials.

GI-POF experiments have been undertaken based on a material called perfluorinated plastic (PF). PF fibres could have an attenuation as low as 1 dB/km at 1300 nm with a fibre of about 750 μm diameter and a 400 μm core. PF is still very expensive and nobody has achieved a mass production version of this fibre although many laboratory experiments have given encouraging results.

Today plastic fibres are mostly used for illumination or very short distance communication systems, such as in a car. The main advantage of plastic fibre is ease of connectorisation but it has yet to prove itself in terms of cost, bandwidth, attenuation and long-term thermal stability. The ATM forum has approved plastic fibre for 155 Mb/s over 50 m of plastic fibre. This is seen as more of a marketing exercise by the plastic fibre lobby than a practical solution, especially when enhanced category 5 copper cable can offer 1000 Mb/s over 100 m.

# 12

## Optical cable technology — cable

## 12.1 Introduction

The purpose of an optical cable is to protect the optical fibre so it can survive the installation process and long-term life in its final environment. The nature of the cable environment is the principal factor in determining optical cable design and in this section we will review all major cable styles and their most appropriate use in cost-effective project planning. The steps in selecting an optical cable are as follows:

- What kind of fibre is required and to what specification?
- How many fibres are required for the link? (Allowing at least 50% extra fibres on top of minimum day-one requirements)
- What is the environment the cable will be installed in?
- What is the most cost-effective cable design to meet the preceding three requirements? Note that full installed cost must be taken into account, not just basic cable cost.

### 12.1.1 Cable environments

Least harsh    Office based in trunking and dado rail.
Riser, ceiling and underfloor cable trays.
External tray work but not directly exposed to the elements.

External tray work directly exposed.

Aerial mounted on support, catenary or messenger wire.

Aerial self supporting.

Underground flooded cable ducts.

Directly buried in the ground.

Exposed tray work in hazardous areas, e.g. oil rigs, ships, mines, petrochem works.

Directly buried in hazardous area, e.g. oil refineries.

Most harsh    Undersea, ocean bed.

The cable design has to be most appropriate for the environment. A small, light cable for short office runs would not survive for very long on the deck of an oilrig. Similarly a double armoured transoceanic cable would be too big and expensive to be run around an office.

Optical cable designs fall into a number of categories but there are two principal groups, which are, tight buffered and loose tube. Tight buffered fibres have a layer of mechanical protection in intimate contact with the fibre. Loose tube cables have tubes containing optical fibres that are loosely contained within. There are some further variants of the latter such as slotted core and ribbon fibre. Tight buffered fibres, with their rugged nature, are preferred by many installers for their easier on-site termination. Loose tube cables give better mechanical protection for the fibres as the fibre is physically decoupled from most of the stress placed on the cable. Loose tube cables also give a better packing density of fibres for the same cross sectional area and thus generally have a lower cost per cabled fibre. Other cabling options to be considered are various armouring and speciality sheathing variants.

## 12.2    Tight buffered optical fibre

Tight buffered optical fibre has extra layers of protection added directly onto the 250 µm primary coating to build up the fibre diameter to around 900 µm, or 0.9 mm diameter. This style of fibre is viewed as being more rugged and easy to handle and best suited to on-site termination of the fibre, i.e. putting a connector directly onto the end of the fibre on-site rather than splicing on a factory-terminated tail cable.

Design options include extruding a PVC layer directly onto the primary coating and a two stage coating which involves adding a soft-silicon layer and then an outer layer of hard nylon to bring the diameter up to the customary 900 µm.

Tight buffered fibre on its own may be used to make up short tail cables i.e. a length of fibre with a connector on one end only, for use within patch panels or test equipment leads. Generally, however, the fibre needs be incorporated into a larger cable construction to be useful.

### 12.2.1   Patchcord

The simplest optical cable is a single tight buffered fibre contained within annular aramid yarns and held together within an outer sheath. The aramid yarns act as the strength member for the cable. The strength member is a concept common to all optical cables as the cable must be designed so that the optical fibre sees as little stress as possible. The strength member can be centrally located or distributed about the cable and can be made of such material as aramid yarn (more commonly known by their brand names, Kevlar® (Dupont) and Twaron® (Akzo)), glass yarn, resin bonded glass or even steel wire.

The single fibre, tight buffered cable is made to a diameter of between 2.5 and 3.2 mm (typically 2.8 mm). All commonly used optical connectors are made with a back-end designed to accommodate such a size of cable. The back-shell of the connector body will terminate and grip onto the aramid yarn so that the yarn sees all of the tensile and other loads before the optical fibre.

Such a small and simple cable would only be used for patchleads and test leads. A variant is to extrude two such cables together to form a duplex structure, also known as zipcord, shotgun or figure-

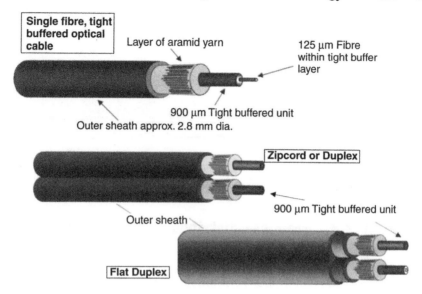

**Fig. 12.1** Tight buffered optical fibres.

of-eight style. A yet more rugged construction is to extrude another sheath over the original two tight-buffered units to form a flat duplex cable which has been used in fibre-to-the-desk installations. Figure 12.1 demonstrates these three styles.

## 12.2.2   Breakout cable

A very rugged cable style for indoor use is known as a breakout cable. This consists of any number of the 2.8 mm units laid up together and protected with an outer sheath. This cable is sometimes used without patchpanels and is connected directly between active equipment. It can be seen from Fig. 12.2 that this style contains a high material and labour content and is thus expensive. Its use in most countries has declined in favour of the cheaper and simpler premises distribution style of tight buffered optical cable.

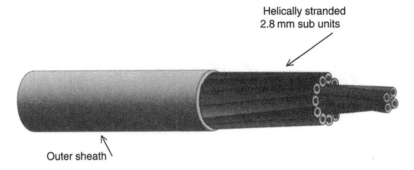

Helically stranded
2.8 mm sub units

Outer sheath

**Fig. 12.2** Breakout optical cable.

## 12.2.3   Premises distribution tight buffered optical cable

Premises distribution, sometimes referred to as light duty style, is the name given to a class of optical cables consisting of a group of 900-μm tight buffered fibres all sharing a common sheath. The fibres, which will be differently coloured for easy identification, are either laid in straight and mixed with aramid yarn strength members or else stranded around a central strength member for the larger fibre count designs.

This style of cable is very popular for building backbone and riser cables because of its ease of termination and relatively low cost. Eight, twelve and sixteen fibre variants of this cable are the most often specified products for building backbone applications. Figure 12.3 shows a six fibre premises distribution cable.

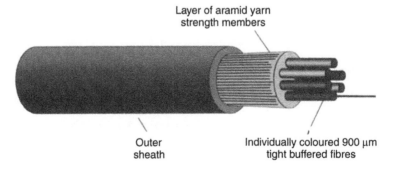

Layer of aramid yarn
strength members

Outer
sheath

Individually coloured 900 μm
tight buffered fibres

**Fig. 12.3** Premises distribution cable.

## 12.2.4   Indoor, outdoor and universal cables

It is worth returning at this point to the issue of cable installation environments. Cables designed for use solely indoors, typically office environments, are mostly concerned about maintaining low flammability requirements. Materials such as PVC, and low flammability, zero halogen materials are specified for this application. Unfortunately these materials have a low resistance to fluids and ultra violet light so they cannot be used outdoors.

Cables used outdoors are concerned entirely with the effects of the weather and waterlogged ducts or trenches. An outdoor cable must be waterproof, resistant to UV light and have a wide operating temperature range. The best materials for this application are polyethylene with petroleum gel filling. In addition, the polyethylene sheath is usually carbon loaded to improve UV resistance and that is why most external cables are black. Unfortunately polyethylene and petroleum gel are very flammable, and most countries prohibit their use within buildings, insisting upon their termination to indoor grade cables within 15 m of entering the building. Failing that, the cables must be entirely contained within metal conduit.

Forming a transition joint between internal and external cables has never been seen as much of an issue with copper cables. Copper pairs are easily and quickly punched down onto blocks and there is anyway the need to break down the high pair count external cable to the smaller pair count riser cables. Added to this is the common requirement to insert over-voltage protection devices at the entrance facility as well.

Putting a transition joint in an optical cable is very labour intensive and expensive. The external cable route may only be a few tens of metres if buildings on the same campus are being connected and there is no need for overvoltage protection.

For this reason universal or indoor/outdoor grades of cable have become popular in the premises cabling industry. A universal cable has sufficient low flammability (and often zero halogen content) to meet national requirements yet maintains sufficient water and UV resistance to survive installation within flooded underground ducts. A higher performance material is required for the sheath to achieve these requirements yet the extra cost is easily absorbed within the lower labour costs of having only one cable to install on a site. On a typical industrial or academic park it is very common for cables to have to run tens of metres within a building and then tens of metres in between buildings. It is most cost effective in these instances to use just one common grade of universal cable.

In some designs a thick layer of low flammability material is used for the sheath and other designs use an inner layer of polyethylene with an outer sheath of low flammability, zero halogen material (polyethylene is intrinsically zero halogen.)

To remove the need for water blocking petroleum gel a newer method is to use water-swellable tapes and threads. Apart from the flammability issues with petroleum gel it is also unpopular with installers because of the mess and time taken to clean away all the gel before termination or splicing can take place. The threads and tapes swell considerably when in contact with water and this prevents water from flowing down the interstitial spaces within the cable.

Most of the cables discussed in this chapter could be made as indoor, outdoor or universal grades but because of its popularity, particular emphasis is given to the premises distribution style for the latest generation of universal grade optical cable.

There is an American marking scheme for optical cables which dictates in which part of a building they may be installed. This is shown in Table 12.1.

Table 12.1 American optical cable marking scheme

| Cable title | Marking | Test method |
| --- | --- | --- |
| Conductive optical fiber cable | OFC | General purpose UL 1581 |
| Non conductive optical fiber cable | OFN | General purpose UL 1581 |
| Conductive riser | OFCR | Riser UL 1666 |
| Non conductive riser | OFNR | Riser UL 1666 |
| Conductive plenum | OFCP | Plenum UL 910 |
| Non conductive plenum | OFNP | Plenum UL 910 |

# 12.3    Loose tube optical cable

The other major family of optical cables is known as loose tube. The tube is typically two to three millimetres in diameter, gel filled and made of a material like PBT (polybutylene terephthalate). Typically twelve fibres go in each tube but 96 fibres have been achieved.

The individual fibres are coloured to ease identification when splicing or terminating, but as the fibres are only 250 μm diameter, primary coated, the human eye could not readily differentiate between more than 12 colours. In cases where there are more than twelve fibres in a tube they are bunched together in groups by different coloured threads.

The main advantage of loose tube cabling is that the fibre is mechanically decoupled from the cable in the normal course of installation and operational use. As the tube is bent or stretched the fibre has some freedom to take up the position of least stress and strain.

Loose tube cables offer the highest possible packing density of fibres per unit area with hundreds or even thousands of optical fibres being contained in cables 20–30 mm in diameter. The majority of external cables are of the loose tube design due to the superior strength they exhibit and their high fibre-packing density.

## 12.3.1    Multiple loose tubes

For higher fibre counts and for maximum strength the multiple loose tube cables are the most suitable. A typical design consists of a central strength member, which can be steel, aramid yarn, or resin bonded glass (RBG) and around which is stranded six or more fibre tubes or fillers. The tubes are stranded around the central strength member to equalise the loading on the fibres. If the tubes were laid in straight, i.e. parallel to the central strength member, then when the cable was bent the fibres on the outside track would be under tension and the fibres on the inside track would be under compression.

The large central strength member makes the cable ideal for pulling into long underground cable ducts. A cable with 6 tubes of 12 fibres each packs 72 fibres within a very small space. By varying the fibre count in the tubes, exactly the same cable carcass design can handle very different quantities of optical fibres. For lower fibre counts, some of the tubes may be replaced with simple plastic fillers just to keep the design circular. It is also necessary to have a method of identification for each fibre within the cable. Each fibre within each tube is coloured, as already discussed, and so each tube must also be coloured as well or else just two tubes are coloured in what's known as a marker-reference system.

Multiple loose tubes, with their high packing density, are very popular with PTTs (public telecom operators) but a high fibre count cable also needs a sophisticated cable management and termination regime, especially when bend-sensitive singlemode fibres are being used.

Loose tube cables can be non-metallic, metallic and may also be armoured for more severe environments such as direct burial. Totally non-metallic optical cables are more popular in datacom applications whereas longer distance telecom applications are still content to use metallic construction optical cables.

A typical non-metallic cable would contain a non-metallic central strength member such as RBG. Stranded around it could be six fibre-tubes, gel-filled. The interstitial spaces between the tubes may also be flooded with gel if it is an outdoor cable in which case there would be a layer of paper tape to hold the whole thing together while it went through the next manufacturing process, i.e. extrusion of the final sheath. The final sheath would be polyethylene for an external grade cable.

If it was a universal cable then the interstitial gel filling would be replaced with water-swellable tapes and threads and the outer sheath would be low-flammability, zero-halogen compound. Figure 12.4 illustrates a typical multiple loose tube, non-metallic duct cable. Figure 12.5 shows a metallic duct cable with a central strength member of stranded steel wire and an aluminium tape moisture barrier.

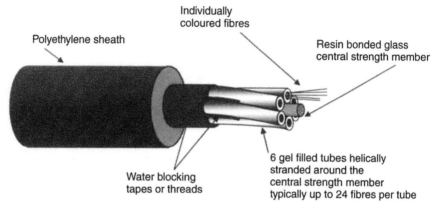

**Fig. 12.4** Non-metallic loose tube cable.

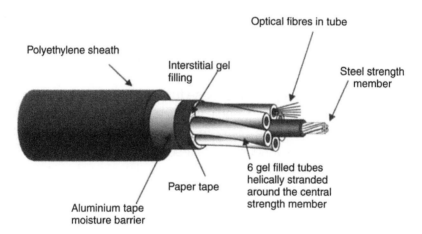

**Fig. 12.5** Metallic loose tube cable.

## 12.3.2   Single loose tube optical cable

A low cost alternative to multiple loose tube designs is to have just one fibre tube in the middle of the cable with annular strength members around it and an overall sheath. Some manufacturers also refer to this style as unitube.

The tube is generally the same as in multiple loose tubes but may be slightly larger. The cable can be made in external duct style or uni-versal grade. Unitube is not as strong as multiple loose tube with its

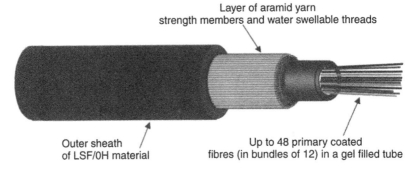

**Fig. 12.6** Unitube optical cable.

big central strength member but unitube is probably the lowest cost design available and is suitable for shorter campus links. Figure 12.6 demonstrates the unitube style.

### 12.3.3   Slotted core optical cables

An alternative style to the loose tube construction is known as slotted core. This design of cable uses a shaped, extruded centrepiece, which contains one or more slots. In cross section the shape may look like a cogwheel or a Maltese cross (see Fig. 12.7).

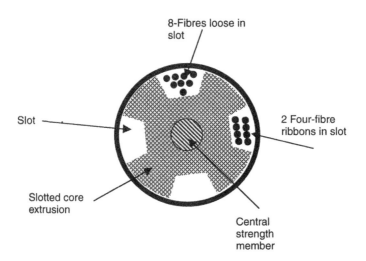

**Fig. 12.7** Slotted core optical cable in cross-section.

The group of fibres sits in the slots in the same way as if they were in tubes. The slots spiral around the centrepiece to serve the same purpose as the tube stranding. Some manufacturers and users believe this to be a cheaper design than conventional loose tube but it has not gained acceptance in many countries as the first design of choice.

The shape of the slotted core ideally fits in with the use of ribbon or ribbonised fibres.

## 12.3.4   Ribbon fibre

Primary coated optical fibres can be bonded together to form a ribbon construction. They can be edge bonded, Fig. 12.8, or encapsulated, Fig. 12.9. The ribbon is typically made up of 4,8,12 or 16 fibres.

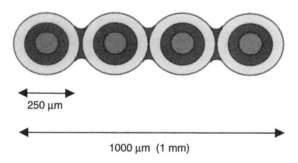

250 μm

1000 μm  (1 mm)

**Fig. 12.8** Edge bonded optical ribbon.

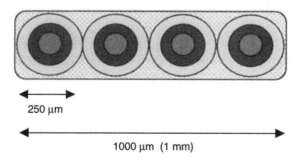

250 μm

1000 μm  (1 mm)

**Fig. 12.9** Encapsulated ribbon fibre.

Ribbons can themselves be encapsulated to form a simple premises cable or they can sit in a tube or they can be stacked up in the slot of a slotted core cable. Each fibre is individually coloured and if there is more than one ribbon in a tube or slot then an identity number can be ink-jet printed onto the flat face of the ribbon to aid identification by the installer.

The main advantage of ribbon fibres is that they can be mass spliced and terminated. To join individual optical fibres together they must be fusion spliced or held together within a mechanical splice. Today ribbon fusion splicers and ribbon mechanical splices can join two optical fibre ribbons together in just one operation. If one were using a 12-fibre ribbon this represents a 12-fold increase in splicing productivity compared to using individual fibres.

When a single fibre is connectorised one has the choice of splicing on a factory-made tail cable or directly connectorising the fibre itself on-site. Ribbon fibres can be spliced onto factory made 4, 8, 12 or 16 way ribbon tails or alternatively a ribbon or array connector, such as the MT, can be directly terminated on-site. There is once again a four to sixteen increase in termination productivity when using ribbon fibre.

Ribbon optical cables are slightly more expensive to make as there is one more process involved, i.e. ribbonising the individual optical fibres, but in fibre-rich installations, as seen in metropolitan area networks and CATV, ribbon cable can prove to be the most cost-effective cable design choice.

# 12.4   Armouring styles

If a cable is not being installed in the relatively benign environment of an office or even an external cable duct then it probably needs to be armoured. Armouring gives the cable much greater compressive strength and prevents attacks from rodents and insects.

## 12.4.1   Steel wire armour

Steel wire armour (SWA) is the strongest of the armouring options. It involves stranding galvanised steel wires around the cable and then

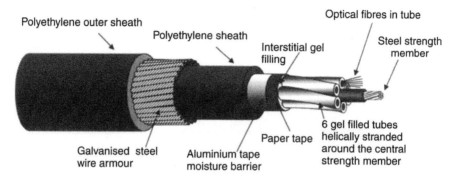

**Fig. 12.10** Steel wire armoured cable.

extruding over it a final sheath. Some very heavy-duty cables have two layers of steel wire wound on in opposite directions. Figure 12.10 shows a cable typical of heavy-duty use such as direct burial in the ground in a cable trench.

The design of this cable promises a very long life. It starts in the centre with a galvanised steel strength member. Stranded around it are six gel-filled tubes and around those is a layer of paper tape (to hold the whole construction together before it goes onto the next process), a layer of bonded aluminium foil to offer a completely water-tight seal and an inner sheath of polyethylene. All the cable interstices are also flooded with petroleum gel. The galvanised steel wires are applied next and then a final outer sheath of a material such as PVC or polyethylene.

## 12.4.2   Corrugated steel tape

A lighter, more flexible and lower cost alternative to steel wire is corrugated steel tape. The tape appears quite thin when seen in its original form, but when corrugated and oversheathed with polyethylene it provides a very strong layer of protection. This construction of cable is especially popular to resist rodent attack. Figure 12.11 demonstrates this style and shows a non-metallic inner core.

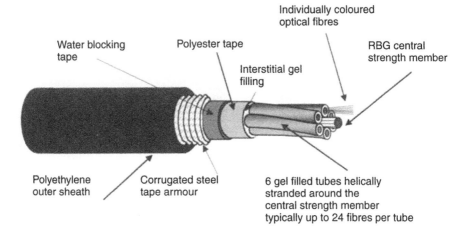

**Fig. 12.11** Corrugated steel tape armoured cable.

### 12.4.3   Steel wire braid

A very flexible form of armour is galvanised steel wire braid. The braid is applied over the cable and then an outer sheath is extruded over that. The flexibility and strength of this design makes it suitable for external cable trays where strength and the ability to negotiate tight corners is a requirement, e.g. on oil rigs and oil refinery pipe and tray work.

### 12.4.4   Non-metallic armour

The three options described above all add metal to the cable which may have started out as a totally non-metallic design. Some non-metallic armouring options are:

- *Glass yarn.* A layer of glass yarn applied under the outer sheath gives a useful degree of added strength to the cable. It is also very difficult, and unpleasant for rodents to gnaw through the cable.
- *Nylon.* Nylon is very hard and slippery and when added over a sheath of high-density polyethylene makes the cable very tough and almost too slippery for small rodents to get a grip onto with

their teeth. Nylon has the added benefit of being oil and chemical resistant.

# 12.5   Special optical cables

There are many projects and applications where special designs of cables and even specialised fibres are required. Some of the applications and design options are summarised here.

- *Copper and fibre composite cables.* Copper wires and optical fibres can be mixed in the same cable. The copper wires can be used for carrying power to remote equipment or to carry low speed data or switching signals. One design places optical fibres in the centre of the cable and puts a layer of twisted copper pairs around the outside. An application for this style is in surveillance cameras on highways. The optical fibres bring the video signal back from the camera to the operations centre, and the twisted pairs take the low speed control signals such as pan, tilt, zoom etc. to the cameras.
- *Lead sheathed cable.* In oil refineries it is often a requirement that cables carrying process-critical information must be lead sheathed. This is because hot crude oil will eventually dissolve any kind of plastic, no matter how exotic its make-up. Lead is totally impervious to oil and will guarantee the signal integrity forever.
- *Radiation resistant fibre.* Optical fibre goes dark in the presence of ionising radiation, i.e. the attenuation goes up to unacceptable levels and the circuit will cease to work. The effect is more pronounced in multimode fibres. Some military and nuclear power/processing installations therefore specify radiation resistant fibre. Another option is of course to use the lead-sheathed cable.
- *Oil/chemical resistant cables.* There is a wide variety of materials that have a greater resistance to solvents than can be obtained from day-to-day materials such as PVC and polyethylene. Nylon is one of the cheapest methods of obtaining oil and chemical resistance. Applications such as coalmines and oilrigs need

cables that are resistant to solvents ranging from seawater to hydraulic fluid. A resistance to ozone and UV light may also be needed, as is a requirement to maintain flexibility across a wide range of temperatures. A useful specification that can be used here is the British Naval Engineering Specification DEF-STAN 61-12 part 31 (formerly known as NES 518).

- *Field deployable cables.* In some cases it is necessary to roll out an optical cable, use it and then roll it up again and take it away. Conventional optical cables, their connectors and even the cable drums they come on are all totally unsuitable for this application. Cables, connectors and the special drums that can be used for this requirement are often called field deployable. They are used by the military for temporary communications, e.g. for connecting anti-aircraft missile launchers and their fire control unit, and for television companies that need to lay out cabling for an outside broadcast event, such as a sporting fixture, and need to recover them again afterwards. The cables involved have to be very strong and flexible over a wide temperature range and the most popular construction for this is four tight buffered fibres within a bed of aramid yarn all contained within a polyurethane outer sheath. Special connectors, called expanded beam connectors, are also required.

# 12.6   Blown fibre

Blown fibre, or air blown fibre, is not to be confused with air blown cables. An air blown cable is a relatively conventional optical or copper cable that is assisted in its passage through a cable duct by the action of air pumped alongside it. Blown fibre is of course similar in concept but consists of specially coated fibres or groups of fibres blown into small plastic ducts.

The purpose of blown fibre is the reduction of initial installation costs. Rather than installing large quantities of dark fibre i.e. spare unconnected fibre for future use, blown fibre allows for the installation of low-cost plastic ducting that can have optical fibre blown into it at a future date as and when it is needed. Blown fibre clearly becomes more economic the larger the project. Small installations requiring a single eight-fibre backbone, for example, may be better off with a conventional optical cable, but a large site with potentially thousands of fibre links worth hundreds of thousands of pounds can save over 65% of initial cabling costs with blown fibre. The final costs, once the blown fibre is added, are roughly the same as conventional cable or even slightly higher. But the main point is the saving of large amounts of capital in the early years of a project. To summarise, the advantages of blown fibre are:

- The ability to defer costs into the future.
- The ability to blow out old fibre and re-use the ducts.
- The ability to defer difficult decisions about fibre type into the future, e.g. singlemode fibre, laser-launch optimised 50/125 etc.
- The ability to repair the ducts more easily than conventional optical cable.
- The ability to add fibre without any disruption to the office working environment.

The disadvantages are:

- Limited number of suppliers.
- No cost advantage for small projects.
- The need to record and protect the ducts so they can be used in the future.

## 12.6.1   Background

Blown fibre was pioneered by British Telecom in the 1980s and is still extensively used by BT in their local network. BT has licensed several manufacturers around the world to produce their own versions.

## 12.6.2   Blown fibre ducts

The blown fibre ducts are plastic tubes around five to eight millimetres in diameter. They are supplied singly or in groups, typically four or seven or even more and packaged in an outer sheath. Waterproof and armoured versions are available for outdoor use and flame retardant versions for indoor use.

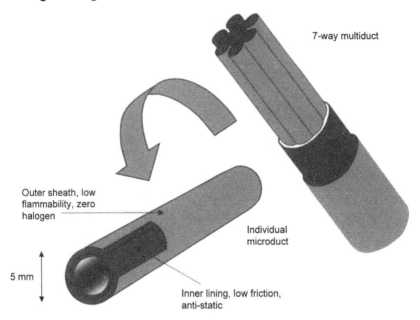

**Fig. 12.12** Blown fibre ducts.

The build up of static electricity is the biggest problem encountered by blown fibre ducts. A plastic coated fibre blown against the inside of a plastic tube would end up sticking to it due to static after a few tens of metres. The duct therefore has an inner lining of carbon-loaded polyethylene which is sufficiently conductive to leak away static electricity.

The ducts are easily joined together, e.g. from outdoor grade to indoor grade, by the use of push-fit pneumatic connectors. Figure 12.12 shows a group of blown fibre ducts.

### 12.6.3   Blowable fibre

The optical fibre used is standard, primary coated fibre. Applied to it is a low-friction, anti-static coating to raise its diameter up to around 500 μm, or 0.5 mm. Alternatively a bundle of fibres, such as four or seven are oversheathed with a low friction material. Optical ribbons can also be blown into ducts. The coating material is generally a UV cured acrylate containing PTFE chips or tiny glass beads and this combination gives the desired qualities of low static build-up, low friction and a high aerodynamic drag factor.

### 12.6.4    Blowing equipment

Either clean dry air or any other clean dry gas, such as nitrogen, can be used. Compressed air or gas can be supplied in bottles or compressed air can be generated locally by use of a compressor. The compressor does not have to be particularly special but it must have the capability to clean and dry the air. Flow rates vary but obviously a larger compressor is required to blow fibre 2 km rather than 50 m.

A machine is needed to guide the fibres into the duct and allow the air to flow over the surface of the fibre and pick it up. The blowing machine is based on a simple tractor feed mechanism that pushes the fibres into the ducting. The blowing machine needs to be able to control fibre speed, air flow, keep count of the fibre quantity blown in and be able to detect fibre blockages and automatically stop.

### 12.6.5    Termination and testing

Once stripped of their special coating the fibres can be spliced or connectorised just like any other kind of fibre. The fibre terminations can be organised in patch panels in just the same way as conventional optical cables. A major difference with blown fibre ducts compared to optical cable is the need to pressure test the ducting before the fibre is blown in. This is to ensure that the duct has not been crushed, kinked or punctured during installation. The duct has a similar size and strength as category 5 copper cable but it still needs to be installed with some care.

The pressure test consists of blowing a steel ball from one end of the duct to the other. At the remote end is a special valve that catches the ball. If the ball arrives one can safely assume that the duct is not damaged. To check the pressure integrity however the ball closes off a valve which then allows the air pressure to rise in the duct, if the compressor is still turned on. The equipment allows the pressure to rise to its maximum working pressure of ten bar, and if the pressure holds then one can conclude that the duct has no punctures or tears in it. The duct is then depressurised and sealed at both ends. This is important to prevent dust, water and insects from entering into the duct. The sealed duct can then have fibre blown into it at any time in the future, however if some considerable time has elapsed then it would be wise to do another pressure check before attempting to blow fibre.

## 12.6.6   Blown fibre capabilities

Blown fibre bundles are optimised for long straight distances as may be encountered in external routes. Two kilometres may be blown in one go and one method even allows fibre to be picked up and blown into the next section.

For internal cable routes, as usually encountered in LANs, numerous tight bends that need to be negotiated are more of a problem than achieving long route distances. Individual blown fibres are better at taking numerous bends and one method allows up to eight optical fibres to be blown along a route containing three hundred 25 mm bends.

The individual fibre blowing method can blow fibre 500 metres through the 5 mm duct and over 1000 m of the 8 mm duct. Vertical rises of at least 300 m are also possible.

With both ends disconnected, and using the same compressor, the optical fibres can be easily blown out and the duct re-used. Additional fibres cannot be added to an already populated duct as they would intertwine and be unable to progress. When blown fibre is used for backbone cabling it is common to use a seven tube bundle, which is potentially a 56-fibre cable, and to blow in four or eight fibres for immediate use, leaving the capacity of the other six tubes available for future expansion.

# 13

# Optical cable technology — components

## 13.1  Introduction

Optical cable cannot be used on its own to create a viable communications network. The cable has to be terminated with connectors, joined by splices and all have to be protected and managed within wall and rack mounted patchpanels, joints and other ancillary equipment. Telecommunications optical networks use more esoteric components such as optical splitters, wavelength division multiplexers and erbium doped optical amplifiers.

## 13.2  Optical connectors

Every optical fibre has to be terminated with a connector to allow it to plug into the active communications equipment. Most optical connectors are of the ferrule type; that is, they consist of a precision made ceramic or polymer tube that has an internal diameter of a few microns more than the cladding of the fibre. In most cases this is approximately 125 μm. The rest of the body of the connector is designed to support and protect the ferrule and the fibre that goes into the back of it. Two of these connectors are designed to mate together through an adapter or uniter so that the polished end faces

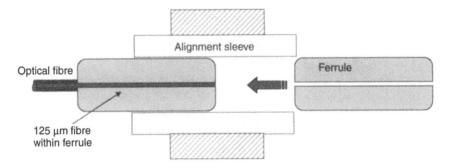

**Fig. 13.1** Ferrule optical connector.

of the fibre cores are closely aligned, allowing the light to cross from one core to another with an acceptable loss en-route. Figure 13.1 shows a simple ferrule connector.

There is another style called expanded beam. A spherical lens is aligned to the end of the fibre. This lens expands the beam and sends it out in a nearly parallel beam across to the other connector. Here the light meets another spherical lens that focuses the light back down into the other fibre core. This style of connector is popular with military, tactical and field deployable cables because the expanded beam cuts down the tolerance required in aligning the cores and is more forgiving of dirt or water contamination. The spherical lens is covered with a flat glass plate to allow for an easy wipe clean and the connector body is hermaphroditic, i.e. any two identical connector shells can join together. Figure 13.2 shows the expanded beam concept.

There are numerous body styles for optical connectors and seven methods of termination.

## 13.2.1   Termination methods

- *Heat cured epoxy*. Sometimes called 'pot-and-polish'. The fibre is first stripped of its primary coating down to the 125 μm cladding layer. Adhesive is injected into the back of the ferrule with a special hypodermic needle. The fibre is then pushed into the back of the ferrule until it protrudes from the front face. The connector is then

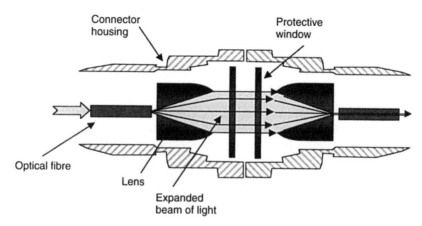

**Fig. 13.2** Expanded beam connector.

placed in a specially constructed oven for about 10 minutes. The adhesive cures and sets permanently after this time. The protruding fibre end is cleaved off with a scribe and then the connector end-face is polished on a series of abrasive papers. The different grades of abrasive paper have a smaller and smaller grit size until the end of the fibre is perfectly smooth and polished and level with the face of the connector. The fibre end is then inspected with a special microscope, such as a Priorscope, and if it looks optically good it is passed as fit for purpose. The connector may also be optically tested by mating it with a known test connector and measuring the attenuation across the two end faces with a power meter and light source. The whole process, excluding the power meter test, takes about 12–15 minutes per connector. The end result however is a permanent and usually high quality optical termination with good thermal stability.

- *Cold-cure or anaerobic*. The fibre is stripped and an adhesive is injected into the back of the ferrule. But this time the adhesive is part of a two component system. The adhesive is only cured when it comes into contact with an activator. After preparing the fibre and injecting the adhesive into the ferrule, the bare fibre is dipped into the liquid activator. Before the activator can evaporate the fibre is pushed into the back of the ferrule. Within 10 seconds or so the activator and adhesive have reacted and set. The protruding fibre end is then cleaved and polished as before. There is no need for oven curing.

- *UV cured*. The fibre and adhesive are prepared as before but it is ultra violet light that cures the adhesive. Once the fibre has been pushed into the back of the ferrule it must be exposed to a strong UV light. After a few tens of seconds the adhesive is cured. The protruding fibre end is then cleaved and polished as before.

- *Hot-melt*. In this method the ferrule is preloaded with an adhesive which softens when heated and then hardens again at room temperature. The ferrule is first heated in a special oven until the adhesive becomes soft enough for the fibre to be pushed through. The connector is left until it has cooled down and the adhesive has set again. The protruding fibre end is then cleaved and polished as before.

- *Crimp and cleave*. A very different method, but one promoted for its speed, is called crimp and cleave. The prepared fibre is pushed into the back of the fibre and simply crimped there. The fibre is then cleaved and may be polished to a small extent. Some questions have been raised about the effect of long term temperature cycling performance on this style. The coefficients of thermal expansion for glass and ceramic (or polymer) are different. As the temperature goes up and down an effect called pistoning may take place if the fibre is not permanently secured in the ferrule, i.e. the fibre will try and retract down the ferrule when it gets colder and will try and protrude beyond the ferrule when it gets hotter.

- *Splicing pre-made tails*. All the above methods are suitable for on-site termination by the installer. A different approach is to terminate the fibre, using any of the above methods, in a clean factory environment, onto a half a metre or more of tight buffered fibre. This is called a tail, a tail cable or a pigtail. The installer will then splice the tail, by fusion or mechanical means, onto the end of the main fibre. This method is preferred when the installation site is very dirty or otherwise difficult to work in and is the main method chosen for telecommunications singlemode fibre systems.

- *Hybrid mechanical splice-connectors*. A mechanical splice is a precision made tube, 126–128 µm internal diameter, that allows two accurately cleaved fibres to be brought into contact with each other. The hybrid connector is a factory-terminated connector with the fibre tail pre-mounted into one half of a mechanical splice. On site the installer merely has to cleave the optical fibre and push it into the other side of the splice where it is retained either by glue or by crimping. This method gives quick and high quality results, but at a price.

## 13.2.2  Optical connector types

There are many different types of connectors in circulation; most with the various termination options described above. Two of the earliest styles were known as SMA and biconnic. In datacommunications and premises cabling the two styles that dominate the market now are the ST (or ST2 as it is often referred) and the SC. The cabling standards, e.g. ISO 11801, state that the SC is the connector of first choice with the ST also recognised. However as the ST is cheaper than the SC, many projects still start off with and stay with the ST.

In telecommunications, where singlemode fibre dominates, the two main connector styles used are the FC-PC and the SC.

Connectors are often described as multimode or singlemode yet both are essentially tubes with 125 μm holes in the middle. The difference is in the tolerance with which the components are made. A multimode connector will typically have a hole size of 127 (+4, −0) μm and a singlemode connector will have a hole size of 126 (+1, −0) μm. The fibre specification is 125 + −2 μm. The ferrule hole size must of course be bigger than the fibre size or you would never get it into the ferrule, but whereas a 62.5/125 fibre could easily cope with a 4 μm misalignment, it would be a disastrous mismatch for an 8 μm singlemode core.

Other connectors in circulation include the duplex FDDI MIC connector and the similar IBM ESCON connector. ESCON connectors are made specifically for the IBM Enterprise Connectivity architecture for connection between mainframes and high-speed peripherals.

Optical connectors end up being recognised with an IEC standard number, but they usually start off life by a manufacturer, or group of manufacturers, taking an idea to the American FOCIS (fiber optic con-

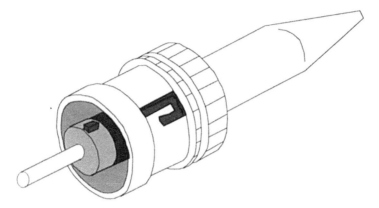

**Fig. 13.3** ST optical connector.

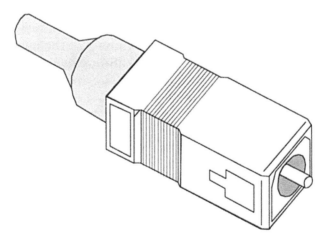

**Fig. 13.4** SC optical connector.

nector intermateability standard) committee, TIA/FO 6.3.4. This committee will assess the viability and technical merits of the connector and vote on its acceptance between the committee members. If a connector is recognised by FOCIS it does not oblige anybody to use it, but other standards bodies such as ISO may then specify it as part of a LAN or cabling system standard. If a connector is seen to be achieving market acceptance, as well as giving some technical/economic benefit then electronic hardware manufacturers will start using it in their equipment.

The ST (Fig. 13.3) and SC (Fig. 13.4) connector have dominated the LAN/premises cabling market for most of the 1990s, but a new

Table 13.1 FOCIS listing of optical connectors

| FOCIS number | Name | Manufacturer/proposer |
|---|---|---|
| 1 | Biconnic | |
| 2 | ST | |
| 3 | SC | |
| 4 | FC | |
| 4a | Angled FC | |
| 5 | MT/MPO | US-Conec |
| 6 | FJ | Panduit |
| 7 | VF | 3M |
| 8 | Mini-MT | Siecor |
| 9 | Mini-mac | Berg |
| 10 | LC | Lucent |
| 11 | SCDC/SCQC | IBM, Siecor |
| 12 | MT-RJ | AMP, Siecor, HP |

generation of connectors is now making an appearance. Most of them fall under the heading of small form factor (SFF). The idea is to make a multi-fibre connector of roughly the same footprint as a copper RJ45 connector. Table 13.1 gives the current line-up of connectors in the FOCIS list with the latest arrivals starting at FOCIS 5.

It would appear that the ISO 11801 and CENELEC 50173 standards are happy to stay with the ST and SC or duplex SC. The new TIA/EIA 568B standard will allow the use of any FOCIS recognised connector but the optical performance and test methods must conform to that standard's requirements, principally an insertion loss of not more than 0.75 dB. The connector manufacturers have still tried to gain acceptance with other relevant standards, namely:

- Fibre Channel    —   VF.
- IEEE 1394b      —   LC.
- ATM Forum      —   VF, LC, MT-RJ, FJ.
- IEEE 802.3z (gigabit Ethernet)  —   duplex SC.
- ISO 11801 —   duplex SC.

Other new connectors are still appearing, such as the MU from NTT and the LX5 from Huber and Suhner.

The MT-RJ (Fig. 13.5) seems to have gained the most market acceptance of all the new connectors so far and numerous items of

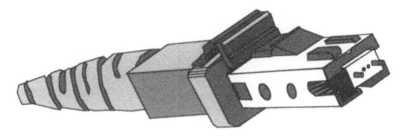

**Fig. 13.5** MT-RJ optical connector.

LAN equipment are now on the market with the MT-RJ connectors fitted. From the perspective of the designer and installer of the premises cabling system the problem remains of which optical connector to use in new installations. The user must carefully weigh up the costs and benefits of the new styles on offer and their support in the future by more than one manufacturer. An interesting market in patchleads does seem to be at least one predictable outcome.

### 13.2.3    Variants and problems of optical connectors

We have seen that connectors can be specified as multimode or singlemode, but this is no more than a measure of manufacturing tolerance. One variant applied to singlemode connectors is an angled front face to reduce return loss. With a perfectly polished and flat end-face, some light will be reflected back from the glass/air interface rather than pass through it on towards the next connector. The light reflected back is called return loss. The ISO 11801 standard limits optical return loss for singlemode connectors to −26 dB or less (note: a bigger number is better!).

Optical energy reflected back into powerful telecommunications lasers could cause them to become unstable. Another problem is ghosting and distortion of the signal by multiple reflections within the optical fibre. This is a particular concern for analogue CATV optical systems.

A method to reduce return loss is to angle the two connector faces so that they do not present absolutely parallel faces to each other, see Fig. 13.6.

Any light reflected back from the angled faces will not be directed

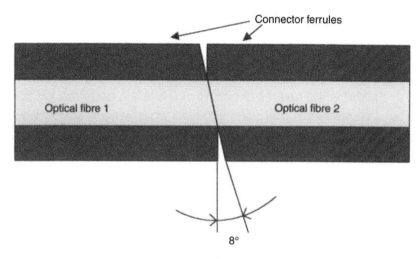

**Fig. 13.6** Angle polished optical connectors.

straight back down the bore of the fibre core as it would with flat polished connectors, and this is sufficient to greatly improve the return loss figure. A face angle of 8° is usually used and special polishing equipment and jigs are required so angle polished connectors are not suitable for field termination.

As yet this does not present a problem for LANs/datacom equipment which has typically used low power LEDs although with the advent of gigabit speeds, lasers and singlemode fibre, it probably will in due course.

Singlemode connectors also tend to use the physical contact method, i.e. the FC-PC, where PC stands for physical contact. The convex end of a PC connector allows the fibres cores to physically touch, whereas a straight, flat polished connector, such as in a multimode ST, will always have a slight airgap between the fibre cores.

## 13.2.4  Optical connector faults

When optical connectors are terminated they are usually inspected by means of a special microscope to gauge the quality of the polished end face. If perfectly done the end of the fibre will be polished smooth and flush with the end face of the connector ferrule. Well-terminated connectors will have an optical insertion loss of typically 0.4 dB, although up to 0.75 dB is allowed in the standards. Poorly finished connectors may have a loss of anything from 2 dB to infinity. Some connector faults, such as partly chipped ends, may appear to have an acceptable loss in one direction but an unacceptable loss when measured in the reverse direction, and this is a good reason why attenuation checks in both directions are worthwhile in optical links.

Optical connectors can often be recovered by the action of repolishing the ends. If this does not improve the optical performance then the connector must be cut off and the process started again. The following is a list of common problems:

- Protruding fibre end from the ferrule body.
- Fibre end 'sunk' back into the ferrule.
- Fibre broken in the ferrule or at the back of the ferrule.
- Fibre end chipped.
- Fibre end badly scratched.

Figure 13.7 gives an example of what some faulty connector ends look like through the microscope.

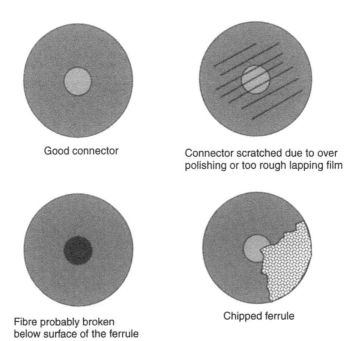

Good connector

Connector scratched due to over polishing or too rough lapping film

Fibre probably broken below surface of the ferrule

Chipped ferrule

**Fig. 13.7** Faulty optical connectors viewed through a microscope.

# 13.3   Other optical termination equipment

Optical fibres, cables and connectors are principal components but there are still others that are needed to fully implement an optical cable system.

## 13.3.1   Optical adapters

Sometimes called uniters or bulkhead adapters, they are required to allow two optical connectors to mate together. The main optical cable will be terminated with an optical connector and the assembly needs to be protected and organised within some form of patch panel. At the front of the patch panel the user needs some form of presentation of the optical connector so that the fibre can be patched into another circuit, or the active transmission equipment, by means of a patchlead.

The front of the patch panel is loaded with optical adapters that are connected to the optical connector on the end of the main cable. An equivalent connector needs to be on the end of the patchlead to plug into the front of the adapter. The adapter is designed to mate particular kinds of connector, so for example an ST connector needs an ST adapter and so on. Once again there are multimode and singlemode options. All adapters are essentially precision made tubes which allow the two ferrules to meet together, and once again the difference between multimode and singlemode adapters is one of tighter manufacturing tolerance for singlemode components. Figure 13.8 shows an SC duplex adapter.

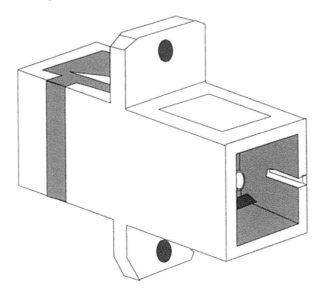

**Fig. 13.8** SC duplex optical adapter.

## 13.3.2   Optical patch panels and joints

As previously noted, the ends of the optical fibres have to be terminated or connectorised to be usefully employed, and those terminations have to be organised and protected so that they present a reliable, useable and manageable presentation to the end user. In datacommunications and LAN cabling the optical cable nearly always ends in a patch panel, which can be rack or wall mounted.

Sometimes optical cables have to be joined together, if runs are very long for example. This is commonplace in telecoms where cable runs are always long and joints need to be made around every 2 km or so. In premises cabling, by the very definition, cable runs are much shorter and so optical joints are not so common.

An optical rack-mounted patchpanel will usually conform to the standard nineteen-inch (483 mm) wide rack-fitting system. The height is measured in 'U', where 1 U is 44 mm, and the depth is variable but needing to be 300–600 mm. Twenty-four optical fibres can usually be terminated in a 1 U panel, but this varies according to the kind of connector used and the point at which the density becomes so great that getting one's fingers around the connectors becomes very difficult. Although rack mounting is most common there is no reason why a patchpanel cannot be wall mounted.

There are two ways to terminate the optical fibre; either direct connectorisation, using one of the methods described earlier, or by splicing on a ready made tail cable. The splice is effected by fusion splicing or mechanical splicing. Either way, the splice and the spare fibre that goes with it need to be organised and protected. The splicing or direct connectorisation process involves a technician working in front of the panel to fit the connector or splice. He/she will need at least a metre of fibre to work with so that the work area can be set up at a realistic distance from the panel. A splice will therefore involve 2 m of fibre as well as the splice itself that needs to be organised and protected. Even with direct connectorisation there will still be a metre of spare fibre to consider. With either method the end result is a piece of fibre with an optical connector on the end. This connector is fitted to the rear-side of the bulkhead adapter that is mounted on the front face of the patchpanel. The end user will then be able to make a continuous optical connection through the fibre network when plugging an optical patchlead into the front of the panel.

A 24 fibre patchpanel may have to accommodate 24 fusion/mechanical splices and up to 48 m of bare fibre. A good patchpanel should have the following features:

- *Cable support/glanding*. As the cable enters the back of the panel there must be a method of securing it so that no mechanical load is ever seen by the fibres, connectors or splices.
- *Splice protection*. Splices should be secured and protected by some device as a cassette system or even a simple row of clips.
- *Spare fibre organiser*. The spare fibre, up to 48 m, must be organised so that the minimum bend radius of the fibre is never infringed and changes and repairs can be made in the future. The fibre cannot just be left to find its own path within the panel. Traditionally, 62.5/125 fibre working at 850 nm has been very forgiving about conditions such as macrobending. Users moving onto singlemode fibre and wavelength division multiplexing in the future will not be so fortunate. The most popular method of controlling

the spare fibre is a pair of plastic crosses that allow the excess fibre to be wound round them.

- *Sliders*. Ideally the rack-mounted patchpanel should be fitted with sliders so that it can slide out and be self-supporting. This makes the installation job much easier with more reliable results.
- *Eye safety*. Now that VCSEL lasers (this is an unfortunate tautology as the 'L' in VCSEL already stands for 'laser', but 'VCSEL lasers' has become a common expression) will be used in nearly all LAN applications, as opposed to only LEDs, then every patchpanel should be marked with the appropriate laser warning symbols denoting the possible class of laser in use. With higher-powered lasers it may be necessary to introduce a method whereby it is impossible or at least difficult for anybody to be in direct line of sight with an optical adapter. If there is no patchlead connected to the adapter then nobody can know if there is any invisible infrared laser radiation being emitted from the adapter. This may be a worse problem for premises cabling than telecoms because cable runs are so much shorter and there will be much more light power reaching the far end. Possible solutions include always capping the optical adapters when not in use or angling the adapters downwards to make direct line-of-sight very difficult.

Large telecommunications networks may have thousands of optical fibres arriving at one location, all singlemode and working at 1310 and 1550 nm. The optical fibre may be standard, dispersion shifted, polarisation maintained or any other manner of speciality. The organisation of the splices, attenuators, splitters, optical amplifiers, wavelength division multiplexers and connectors is critically important or the network just wouldn't be reliable. The passive termination hardware is also a very small percentage of the total and so there is relatively little cost pressure on this item. It seems unfortunate in premises cabling that although people may be spending a million dollars on the cabling infrastructure there is a great resentment about procuring patching equipment concomitant with the sophisticated tasks expected of the cabling system.

### 13.3.3   Optical joints

Optical cables are typically made in 2–4 km lengths. If the cabling run is longer than that then two cables will have to be joined or spliced together. Joints may also be required if large cables have to go round such tight bends that the requisite pulling force would be unacceptable. Broken cables will also need joints inserted in them to repair the cable. There are two kinds of optical joints, in-line and distribution.

- *In-line.* An in-line joint simply joins two optical cables together. Joints usually have to go outdoors and live within inspection chambers buried in the ground or else are mounted up telegraph poles. Rugged mechanical construction and water-resistance are therefore essential design elements for all classes of cable joints. The joint has three main construction elements.
- *Chassis.* A rigid metal chassis is required to secure the cable strength members to and to hold the splicing protection mechanism.
- *Splicing protection mechanism.* The splices and spare fibre must be protected and organised. This is usually done with a cassette-based system.
- *Closure.* A waterproof plastic enclosure must totally cover and protect the splices. The closure may be of heat-shrink material or two halves of a shell that bolt together.

The two cables may enter the closure from opposite ends but more common is the dome-ended joint where both cables enter at one end and an elongated dome closure is placed over the chassis and sealed. Figure 13.9 shows a typical in-line optical cable joint.

- *Distribution joint.* The distribution joint, sometimes called a flexibility point, allows a high fibre count cable to be broken down into smaller cables. This is very useful in networks such as CATV distribution systems or the very large campus cabling systems that may be encountered on big military bases, universities and airports etc. A small campus network using, for example, eight fibres would normally be configured as 'daisy-chaining' from one building to another. That is, the cable is terminated at each

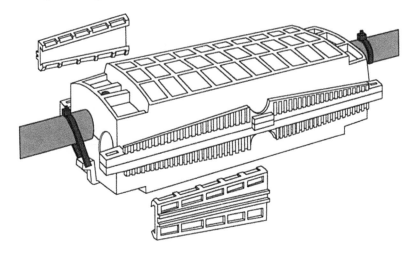

**Fig. 13.9** In-line optical joint.

building and then patched onto the next leg of the cable leaving the building. This method would be totally uneconomic for an 80-fibre backbone cable that had to pass a potentially large number of users, as might be seen in a city-centre wide area network. In such circumstances it would be better to have a distribution joint mounted out in the street and to tap-off a smaller cable, such as four or eight fibre, and drop it off into the users' premises. The distribution joint takes advantage of the construction of multi-loose tube optical cables. An 80-fibre cable may well be constructed of ten tubes with eight fibres per tube. In a distribution joint the main cable is looped into the bottom of a dome-style closure, which has a special oval slot for such a purpose, and coiled inside, but it is not cut. Instead the sheath is removed to expose the fibre tubes. One or more tubes may be cut into and the fibres spliced onto a smaller drop-cable. The splices are protected by a cassette-style mechanism. Up to six such cables may be dropped off in this manner from the main cable at one point. The distribution joint may feature other innovations such as hinged cassettes so those splices can be re-entered in the future without disturbing existing circuits. Figure 13.10 shows the topography of a distribution-joint based network and Fig. 13.11 shows a typical distribution joint.

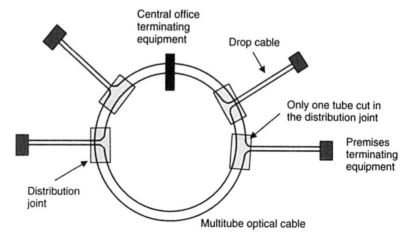

**Fig. 13.10** Topography of an optical distribution network.

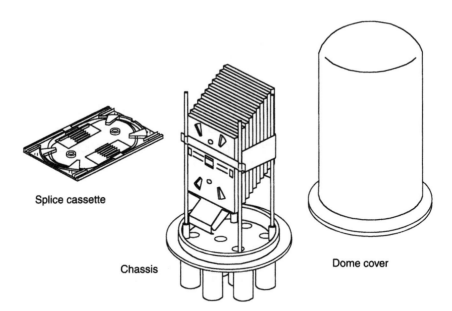

**Fig. 13.11** Typical distribution joint.

# 13.4   Optical splices

When two optical fibres are joined together it is called splicing. Splicing may be effected by mechanical means or by fusion splicing.

## 13.4.1   Mechanical splices

A mechanical splice is a precision made tube that allows two accurately cleaved optical fibres to touch each other and transfer light from core to core with a minimal optical loss.

The optical fibre is first stripped down to the 125 µm cladding layer. An optical fibre cleaver must then cleave it. The cleaver works by scribing a line across the fibre perpendicular to the fibre axis. The fibre is then bent over an anvil and it will normally cleave itself with a clean 90° cut. The 90° cleave is very important as it allows the two fibre end faces to completely touch each other. A lesser or greater angle will not allow the fibre end faces to butt against each other. A chipped or jagged end will also not allow proper contact between the two fibre cores.

Some fibre cleavers are very simple, consisting of no more than a scribing pen with a diamond tipped end. These devices are suitable for cleaving the fibre end protruding from an optical connector ferrule before polishing, but are not accurate enough for splicing. For mechanical and fusion splicing a much more sophisticated device is employed which scribes and cleaves the fibre in a controlled and repeatable manner.

Once stripped, cleaved and cleaned, the fibre is inserted into the mechanical splice tube. The other fibre to be spliced is inserted into the other end until it butts against the first fibre. The accuracy and tolerance of the tube, and the fibre, will determine how well aligned the cores are and hence the resulting optical loss. To improve matters it is common to preload the tube with an index-matching gel. This is a viscous fluid with a refractive index much closer to that of glass than air. It prevents back reflections from the inevitable tiny air gap that will exist between the two fibres and so improves optical loss.

Finally the mechanical splice will be provided with further mechanical means to hold the fibre permanently in place, either by means of clips, adhesive or crimping.

## 13.4.2   Fusion splicing

Fusion splicing works by bringing two prepared optical fibre ends together between two electrodes. An electrical arc is struck between the electrodes and the heat of the arc melts the two ends of the fibres. The fibres are gently pushed together at this stage so they fuse together to form a permanent, low-loss splice.

The fibre ends are prepared in exactly the same manner as for the mechanical splice, i.e. stripped, cleaved and cleaned. The fusion splicer is fitted with precision 'V' grooves that hold the fibres while their end faces are brought together between the electrodes.

Modern fusion splicers are highly automated and sophisticated machines. They are fitted with microscopes and visual displays to allow the operator to manipulate the fibres to bring them together. The more expensive machines do the fibre alignment purely under microprocessor control.

The nature of the electrical arc is also tightly controlled in terms of voltage and current profile and duration. Different programmes are usually available to accommodate multimode and singlemode fibres and even for different manufacturers whose fibres may have slightly different melting properties. Figure 13.12 shows the basis of fusion splicing.

The fusion-spliced fibre will be a permanent and low attenuation joint. Before splicing the two fibres the operator must put a heat shrink splice protector over one of the fibres first. The splice, with its length of exposed 125-μm cladding, is very fragile and needs to be protected. The heatshrink protector is a plastic tube with a stainless steel reinforcing bar. It is slid over the exposed fibre area and completely covers it. Most fusion splicers are fitted with small ovens for such a purpose and they cause the plastic tube to shrink down against the fibre with the steel rod taking the entire mechanical load.

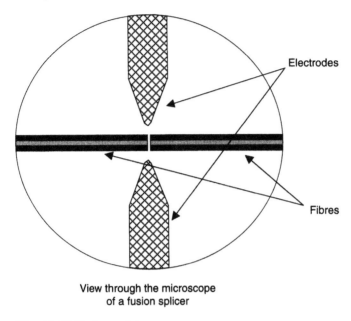

View through the microscope
of a fusion splicer

**Fig. 13.12** Fusion splicing.

## 13.4.3   Mechanical versus fusion splicing

For multimode fibre both mechanical and fusion splicing will give attenuations in the order of 0.2 dB. ISO 11801 allows 0.3 dB for every splice however it is made. Fusion splicing is usually superior for singlemode fibre because of its active alignment systems. Some users are also concerned about the long-term degradation of the index matching gel within mechanical splices, especially in areas of extreme temperature change and low humidity.

The main argument revolves around costs however. A mechanical splice will cost between $10 and $20 and every subsequent splice will also cost the same. There is also an initial layout on the fibre preparation and cleaving kit but this is mostly common to both methods. A fusion splicer will cost between $6000 and $20000 but from then on each splice will only cost pence, i.e. the cost of the splice protector. It can easily be seen therefore that for anybody intending to do more than a few hundred splices a year it will be cheaper to invest in a fusion splicer.

For emergency repairs and remote station repair kits it is easy to make an economic case for mechanical splices, as it is for cable installers who may indeed only be doing a few dozen splices a year.

Latest developments include mass splicing or ribbon splicing. Four, eight or twelve fibre ribbons can be spliced together in one go. This offers huge potential savings for labour costs in fibre-rich installation scenarios. Both fusion and mechanical ribbon splices are available.

# 13.5   Wavelength division multiplexors and other components

There are many other components used in optical communications but they will only be briefly reviewed here because their use in data-communication/LAN networks is still restricted. However as LAN traffic becomes faster we will see datacom and telecom technologies converging together.

The other components include wavelength division multiplexors, splitters, combiners, attenuators, optical amplifiers, lasers and LEDs.

## 13.5.1   Lasers and LEDs

Although we have the beginnings of all optical or photonic switching, all optical communication systems on the market today still need to convert electrical signals into optical signals at the start of the channel and convert the optical signal back into an electrical signal at the end of the channel. There are two families of devices to convert electrical signals into optical signals, light emitting diodes (LEDs) and lasers.

LEDs are cheaper and simpler to make than lasers but they are limited in output power, can only be modulated at modest rates and have a broad spectral width. Lasers have all the opposite properties to LEDs. With LEDs only being able to modulate, i.e. turn on and off, at up to 200 MHz, they are limited to applications in the 100–

155 Mb/s zone. But the vast majority of LEDs currently in use in LANs will still only be working at 10 Mb/s speeds such as 10 baseF.

Lasers, which can modulate up to 10 GHz, have traditionally been the preserve of high-speed telecommunications systems where their high prices have been mostly irrelevant compared to the overall costs of telecommunications projects. Also telecoms exclusively use singlemode fibre which needs a laser to launch light into its very small core. However the LAN environment now includes communications speeds of 100, 155, 622 and 1000 Mb/s and 10-gigabit links are now planned for the next generation of Ethernet. Distance requirements have also crept up as corporations look to extending their LAN into the metropolitan and wide area network. Laser prices have come down but the new driving force in this sector of the market is the VCSEL (vertical surface emitting laser). The VCSEL is a low cost laser made more along the lines of an LED. It is currently available for 850 nm applications but 1300 and even 1550 nm versions are planned. In terms of cost, complexity and performance we can make the following list, in ascending order:

• LED.
• VCSEL.
• Fabry-Pérot laser.
• DFB, distributed feedback laser.

At the other end of the link the device used to turn the optical signal back into an electrical signal is a photodiode. For lower speed applications a device called a PIN is used. At higher speeds, and where more sensitivity is needed at the end of very long links, a device called an avalanche photo diode (APD), is used. The APD uses a high voltage to bias the diode to accelerate and amplify the effect of the photons arriving on the photodiode.

With the problems of chromatic dispersion and the future impact of wavelength division multiplexing, it is desirable to have as much of the signal energy as possible concentrated into the narrowest possible spectrum. LEDs have a wide spectrum with half of their power spread across 30–60 nm. Lasers have a much more concentrated spectrum with a width of 1–6 nm.

Table 13.2 Typical parameters for lasers and LEDs

|  | Centre wavelength (nm) | Spectral width (nm) | Typical output power (dBm) | Max data rate | Detector | Cost |
|---|---|---|---|---|---|---|
| LED | 850 | 50 | −15 | 155 Mb/s | PIN | Low |
| VCSEL | 850 | 1.0 | 0 | 2 Gb/s | PIN | Medium |
| Laser | 1310 1550 | 1.0 | 1 | 10 Gb/s | PIN/APD | High |

First window multimode operation is centred on 850 nm and the sources are LEDs and VCSELs. The detector will be a PIN. Second window multimode operation is centred on 1300 nm and the source can either be an LED, a laser or soon a VCSEL. The detector can be a PIN or APD. Second window singlemode operation is centred on 1310 nm and third window is centred on 1550 nm. Currently all singlemode transmitting devices are lasers. Table 13.2 gives some typical parameters for lasers and LEDs.

## 13.5.2   Erbium doped fibre amplifiers (EDFAs)

Erbium doped fibre amplifiers (EDFAs) and wavelength division multiplexors are complex subjects but we mention them here because with the advent of ten gigabit Ethernet we may well see this technology impinging upon backbone LANs and cabling.

Wavelength division multiplexing (WDM), is a method of putting several data streams, modulated onto light carriers of different wavelength, onto one optical fibre. This can increase the already impressive bandwidth of singlemode fibre by a factor of eighty and in the future by eight hundred!

WDM has evolved into dense wavelength division multiplexing (DWDM), which has a channel spacing of 0.8–1.6 nm (50–100 GHz) with the current technology offering 16–80 such channels in the 1530–1560 nm band. Each channel can take up to 10 Gb/s of traffic. Future systems will use wider bands to get in more channels, possibly up to 800.

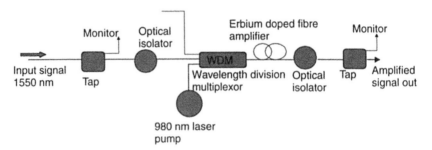

**Fig. 13.13** Erbium doped fibre amplifier system.

Another approach to WDM, for lower cost data transmission systems, is called coarse wavelength division multiplexing (CWDM). Four channels with 20 nm separation are proposed here.

WDM requires amplification of the signal and the subject is generally incorporated under the title of erbium doped fibre amplifiers (EDFAs). An EDFA system has six principle components:

- isolator.
- tap/couplers.
- WDM.
- pump laser.
- photodetectors.
- Erbium doped fibre.

Figure 13.13 gives an indication of how these elements fit together. The erbium-doped fibre is a totally photonic amplifier; there is no need to turn the optical signal back into an electrical signal to regenerate it. The energy to do this comes from the pump laser, which works at 980–1480 nm.

# 14

# Optical cable technology — testing

## 14.1 Introduction

There are three steps required in the successful implementation of an optical fibre cabling system:

- Design it to work.
- Install it correctly.
- Test to determine if the design parameters have been met.

Optical cable links may be tested using an optical time domain reflectometer (OTDR), or by an optical power meter and light source. In practice it is much simpler to use an optical power meter, and just use an OTDR for fault finding. For larger projects the end-user customer may often insist on OTDR trace results to characterise the fibre, but few people really understand the significance of what they see on OTDR traces. For short haul datacom links the optical power meter method is quickest and cheapest and allows for the fastest interpretation of results. All optical test reports must include a table of link loss results either from a power meter or derived from OTDR traces. If a problem does arise the installer must have access to an OTDR for fault finding purposes.

## 14.2   Designing the optical link to work

Every fibre link must have a total link attenuation equal to or less than the total attenuation allowed for that class of link as detailed in ISO 11801. The installer must first calculate what the maximum allowed loss (attenuation) is for the link by adding up the sum total of optical elements in the link, i.e. the cable, connectors and splices, and then measure that link to ensure that the installed results are equal to or less than the calculated attenuation. Table 14.1 gives the ISO 11801 component values to be used in calculations. Table 14.2 gives the link parameter rules from ISO 11801.

The maximum optical attenuation between any two items of optoelectronic transmission equipment should not exceed the link loss required for different optical transmission protocols but should never exceed 11 dB at 850 or 1300 nm. See Table 6.8 in chapter 6 for the attenuation requirements of different optical LANs.

Table 14.1 ISO 11801 component parameter rules

|  | Multimode | | Singlemode | |
| --- | --- | --- | --- | --- |
|  | 850 nm | 1300 nm | 1310 nm | 1550 nm |
| Fibre attenuation dB/km | 3.5 max | 1.0 max** | 1.0 max* | 1.0 max* |
| Fibre bandwidth MHz.km | 200 min | 500 min | n/a | n/a |
| Connector insertion loss dB | 0.75 max | 0.75 max | 0.75 max | 0.75 max |
| Connector return loss dB | 20 min | 20 min | 26 min | 26 min |
| Splice loss dB | 0.3 max | 0.3 max | 0.3 max | 0.3 max |

*ISO 11801 (1995) calls for a singlemode loss of less than 1.0 dB/km. IEEE 802.3z (optical gigabit Ethernet) calls for 0.5 dB/km. The new TIA 568B calls for 1.0 dB/km for internal cables and 0.5 dB/km for external cables. The author recommends a design value of 0.5 dB/km should be used for singlemode fibre.
**This may be increased to 1.5 dB/km in later standards.
*Note:* Bandwidth and return loss are not normal acceptance tests.

| Table 14.2 ISO 11801 link parameter rules | | | | | |
|---|---|---|---|---|---|
| Cabling subsystem | Link length max | Attenuation dB max | | | |
| | | Multimode | | Singlemode | |
| | | 850nm | 1300nm | 1310nm | 1550nm |
| Horizontal | 100m | 2.5 | 2.2 | 2.2 | 2.2 |
| Building backbone | 500m | 3.9 | 2.6 | 2.7 | 2.7 |
| Campus backbone | 1500m | 7.4 | 3.6 | 3.6 | 3.6 |

## 14.2.1  Centralised optical cabling

The model of horizontal, riser and campus cabling described above has been modified to allow for a centralised optical architecture, or COA. This does away with horizontal cross-connects/floor distributors and allows the user to cable optical fibre all the way from the desk back to a central computer room. A cable run of up to 300m is allowed. This concept was first described in *TSB 72, centralized optical fiber guidelines*, October 1995 and *ISO/IEC JTC 1/SC25 PDTR 14763-2 part 2 implementation and operation of customer premises cabling*, November 1997. Figure 14.1 demonstrates this concept.

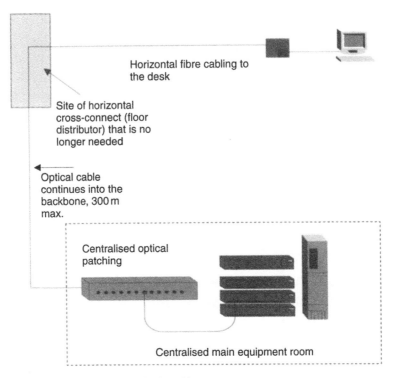

Horizontal fibre cabling to the desk

Site of horizontal cross-connect (floor distributor) that is no longer needed

Optical cable continues into the backbone, 300 m max.

Centralised optical patching

Centralised main equipment room

**Fig. 14.1** Centralised optical architecture.

## 14.2.2   Calculating acceptable optical loss

To calculate the optical loss of an optical cable link it is best to draw a diagram of the link and then ascribe the maximum allowed loss for each item, and then add up each item loss to arrive at the total loss allowed for that link. (Note that the following calculations allow for either 50/125 or 62.5/125 fibre when multimode fibre is referred to). Figure 14.2 gives an example.

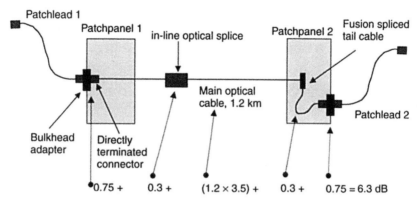

**Fig. 14.2** Optical loss calculation example.

Calculation:

| Connector loss 0.75 dB | Fibre loss 1.2 × 3.5 (i.e. at 850 nm) = 4.2 dB | Connector loss 0.75 dB |
|---|---|---|
| | Splice loss 0.3 dB | Splice loss 0.3 dB |

So to calculate the total allowed attenuation in the above example we add the individual loss components. Remember that the loss of the fibre is different for each wavelength of light used, so one must be specific. In this example we will choose 850 nm first and then 1300 nm.

| | |
|---|---|
| Loss across the first connector pair | 0.75 dB |
| Loss across the cable (1.2 × 3.5) | 4.2 dB |
| Loss across the splice in the main cable | 0.3 dB |
| Loss across the splice in the 2nd patch panel | 0.3 dB |
| Loss across the second connector pair | 0.75 dB |
| Total link loss at 850 nm | 6.3 dB |
| Total link loss at 1300 nm (1.2 × 1) | 3.3 dB |

Is this acceptable under the ISO 11801 design rules?

First of all we must define the link type. As it goes between buildings it is a campus backbone, so the cable run should not be greater than 1500 m. The 1.2 km link depicted is therefore acceptable. Table 14.2 states that the total loss for a multimode campus backbone link should not exceed 7.4 dB at 850 nm and 3.6 dB at 1300 nm. So the figures we have of 6.3 and 3.3 dB respectively are totally acceptable. Finally the designer must ensure that the total loss between the electronic transmission equipment should not exceed 11 dB or any other attenuation budget demanded by any particular protocol. The design of the above example is therefore acceptable within the limitations of ISO 11801.

*Note*: The effect of the patchcords themselves is discounted. This is for two reasons.
- The attenuation of the fibre in a patchcord is negligible.
- The loss across the connector as it goes into the opto-electronic equipment is already accounted for in the launch conditions presumed by the manufacturer.

The designer of the cabling system must be wary of multiple patch panels within the optical link, as it will be seen that the total link attenuation will rapidly rise as patch panels are added in series in the optical circuit.

# 14.3  Testing the optical link

Having calculated the maximum attenuation in the link and determined that it does indeed fall within the rules the next step is to test the link to ensure that the installed cable system has the same or lower attenuation.

## 14.3.1  OTDR

An OTDR (optical time domain reflectometer) is a testing device that can identify faults within an optical cable and state accurately where they are. The OTDR can give a graphical characterisation of the fibre under test. The results however need expert interpretation and so whereas the OTDR is an excellent faultfinding device it is not best suited to simple acceptance tests, especially for multimode fibre.

An OTDR works very much like radar. A short pulse of laser light is launched into the fibre and the light travels until it reaches the end of the fibre where some of it is reflected back to the source. This is known as Fresnel reflection which happens at the interface between materials of differing refractive index, such as air and glass. If we know the refractive index of the fibre under test then we know the speed at which light travels in that particular fibre. After this information is loaded into the OTDR it can then time the point of departure of the light pulse until its return. With a round journey time recorded and knowledge of the refractive index (and hence velocity), the OTDR can accurately work out the overall length of the fibre to within a few metres accuracy.

Apart from this straightforward length measurement capability, the OTDR can also detect the small quantity of light continuously reflected back from small imperfections along the whole length of the fibre, known as Rayleigh scattering. The resulting trace will allow the determination of the attenuation over any length of the fibre or indeed the whole cable run.

Any other discontinuity that doesn't totally break the fibre will also be seen, principally connectors, splices and tight-bend or compression induced attenuation high spots. Figure 14.3 shows a typical OTDR trace.

The capabilities of an OTDR make it ideal for fault finding and characterising long distance telecommunication networks, but the sophistication of the machine and the need to interpret results very often leads to meaningless OTDR traces being presented to the end-user in premises cabling projects.

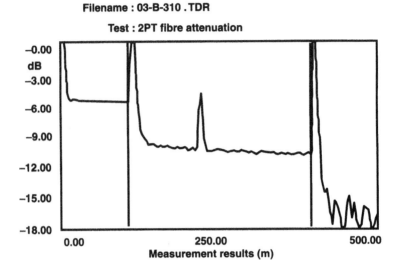

Filename : 03-B-310 .TDR

Test : 2PT fibre attenuation

| Cursor (C) : 400.000 m | Marker (M) : 104.000 m |
|---|---|
| Length diff : 296.000 m | 2 Pnt. loss : 5.217 dB |

Loss/length : 17.624 dB/km

**Fig. 14.3** Typical OTDR trace.

## 14.3.2   OTDR problems

- *Pulse length and dynamic range.* To see detail over a 50 km cable length a powerful laser pulse has to be sent. This equates to the need for a long pulse duration. The length of the pulse of light is equivalent to tens of metres of optical fibre. This gives a low resolution, as events closer to each other than the pulse width will not be seen. There will also be a dead zone in front of the launch device, as the receiver will be closed off until the light pulse has finished.
- *Dead zone 'blinding'.* The amount of light power reflected back from distant events is very small. Local events such as a connector pair a few tens of metres away will reflect back a proportionally much larger signal. This signal can temporarily blind the receiver and it takes a short time to recover. But this recovery period equates to a dead zone forming behind the local event that cannot be seen.

- *Ghosting*. If a lot of power is reflected back from a close event such as a connector pair then some of it may reflect back again from the OTDR towards the connector pair. This can happen several times. The multiple reflected signals will take time to make the round journey and so the OTDR will interpret them as other attenuation events happening at exact multiples of the original event distance.
- *Fibre NA mismatch*. A splice or a connection onto a fibre with a different numerical aperture can reflect back more light than the original part of the fibre. The OTDR will interpret this as a gain in the fibre link rather than an attenuation. The loss of the fibre link can only be accurately determined by measuring it from both directions and averaging the two results.
- *Insufficient averaging*. To see the returning signal the fibre is scanned several times and the results averaged out. This has the effect of cancelling out much of the noise because the noise is random in nature whereas the desired signal will remain the same through every scan. If the OTDR is not allowed enough time to settle down and average the result then the trace will be too mixed in with noise to be readable.

The operator of the OTDR must therefore do the following to obtain meaningful results:

- Load the correct value for the refractive index.
- Set up an appropriate pulse width for the length of fibre being tested.
- Use long enough launch and tail leads, that is extra fibre added at the beginning and end of the real cable link, so that dead zones at both ends of the fibre are removed and there is sufficient resolution to see the far end connector. Figure 14.4 shows the launch-tail lead configuration.
- Allow sufficient time for the OTDR to average out the noise and signal.
- Test from both ends.
- Test at all wavelengths likely to be used, e.g. 850 and 1300 nm.
- Use the OTDR cursors so that the length of fibre under investigation is accurately measured.

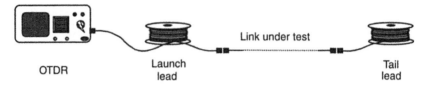

**Fig. 14.4** Test set-up with launch and tail lead.

| Table 14.3 Typical refractive indices of optical fibres | | | |
|---|---|---|---|
| | 850 nm | 1300 nm | 1310 nm |
| 50/125 | 1.481 | 1.476 | |
| 62.5/125 | 1.495 | 1.490 | |
| Singlemode | | | 1.472 |

- Set up the OTDR and use it according to the manufacturers' instructions. Typical refractive index figures are given in Table 14.3.

### 14.3.3 Power meter and light source

This is the simplest and cheapest method of testing a short haul datacoms link, either for multimode or singlemode fibre. The power meter and light source are first calibrated by plugging the light source directly into the meter and setting the meter to read 0 dB. For the rest of the testing session the test lead should not be unplugged from the light source nor should either instrument be switched off. The meter and light source should not be calibrated by connecting the two test leads together using a connector adapter in the middle, because this has the effect of discounting the loss of one of the bulkhead adapters in the cable system under test. However some hand-held cable testers with fibre optic test-set add-ons do require this and so users must always read and follow the manufacturer's instructions. Figure 14.5 shows the calibration set-up and Fig. 14.6 shows the typical test set-up for a working system.

The user must ensure that the light source and power meter are both set to the same wavelength. Note also that the optical fibre is

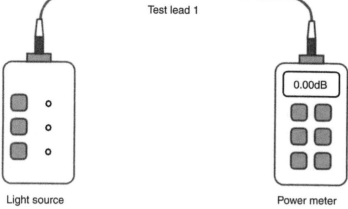

**Fig. 14.5** Optical power meter calibration.

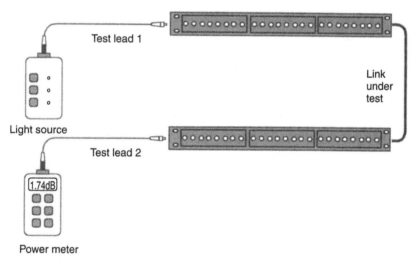

**Fig. 14.6** Optical testing with power meter/light source.

characterised at 850 nm and 1300 nm for multimode, but some test sets do not use exactly these same wavelengths. If this should be a significant problem then table A1 of BS 7718 (*Code of practice for fibre optic cabling*) gives advice on correction factors. The cable should also ideally be tested from both ends. If there is a chipped connector at one end of the link it will allow light to leave it without too much loss, but when light is launched into it from the other direction it will reflect back a large proportion. So testing in one direction only can mask some potential problems.

## 14.3.4   Presenting the results

To be useful the test results must be correctly documented. Hand-held testers for copper cable now store all the test data in an appropriate format but it is not so straightforward for optical results.

OTDR traces can be stored as paper printouts or in electronic form. Power meter results should be tabulated in columns that show the calculated loss for a link against the actual measured loss for both directions and at both wavelengths. The measured attenuation results should always be less than the calculated results or else the link is considered as a fail. A diagram to explain exactly what the test results refer to should accompany the fibre optic test results. The fibre type, length of fibre and positions of connectors and splices must all be identified. Each circuit or optical link must be identified on the diagram. Each fibre in the link must be tested but a multifibre cable may be represented by just one line indicating the number of fibres in it.

# 15

# Cable system design and international standards

## 15.1 Who writes the standards?

### 15.1.1 Introduction

It is important to understand the difference between standards, codes and directives. A standard offers a specification that details how to do something and what physical parameters may be expected as a result. A code or directive is a legal requirement or obligation to do something and usually a national or international standard will be invoked as a measure to determine if the code has been met. The European Union publishes Directives and Statutory Instruments, such as that governing electromagnetic emissions from electrical equipment, and then compares compliance against that directive by invoking EN standards written by CENELEC.

In America the National Electrical Code covers safety issues, such as fire performance, and uses Underwriters Laboratory (UL) tests to demonstrate compliance.

Standards can also cover many different aspects of the premises cabling business. These include:

• Cable system design.
• Component standards.

- Fire performance standards.
- EMC/EMI standards.
- Test standards.
- Earthing, grounding and bonding.
- Cable containment and administration.
- Directives and codes, Europe, America, Canada and Australia.
- Local area network standards.

Standards can also come from different countries. In Europe every country had its own standards writing body, such as the British Standards Institute, NF in France and DIN in Germany. These bodies still exist but national standards have been harmonised to create an open and fair market within the European Union. Most European standards of relevance to the subject of cabling now come from CENELEC, who write all electrical standards for the EU.

In America there are various standards bodies, such as ANSI, TIA and EIA. In Canada there is the CSA (Canadian Standards Association). The European and American standards cover most of the world market for information technology products, but also mentioned in this chapter will be Canadian, Australian and New Zealand standards.

There are also international standards. Principal to structured cabling are ISO (International Standards Organisation), IEC (International Electrotechnical Organisation) and the ITU (International Telecommunications Union).

## 15.1.2  Europe

All European standards of interest come under CENELEC, based in Belgium and set up in 1973 as the officially recognised European standards organisation by the European Commission in Directive 83/189. The CENELEC standards are called European Norms or ENs. Standards not yet published are called preliminary European Norms or prENs.

Every European country still maintains its own national standards body, such as the British Standards Institute in the UK, but CENELEC standards are adopted as national standards where they exist and then, for example, in the UK, 'BS' is placed in front of the 'EN' number. There are some exceptions, such as BS 7718 *The code of practice for fibre optic cabling*, which has no CENELEC equivalent.

In the UK the Fibre Industry Association is a leading trade organisation devoted to LAN based optical fibre and initiated the BS 7718 Code of Practice. The IEE, Institute of Electrical Engineers writes the nationally accepted *Wiring regulations* (also known as BS 7671), which contains safety issues of power cabling, earthing, bonding etc.

OFTEL is the British 'Office of Fair Trading for the Telecommunications Industry'. Their charter states that they are the:

> . . . independent regulatory body with responsibility for ensuring that holders of telecommunications licences comply with their licence conditions. OFTEL maintains and promotes effective competition in the telecommunications industry and promotes the interest of consumers and purchasers of telecommunications services and apparatus in respect of prices, quality and variety.

Oftel's interest in structured cabling is limited to the maintenance of telephone cabling and the approval of electronic equipment which may be connected to national telecommunications networks. Such approval must first be sought from BABT, the British Approvals Board for Telecommunications. Oftel provides a wiring code and this is addressed in the British Standards Institute publication *A guide to cabling in private telecommunications systems* DISC PD1002. There is input to this document from the Telecommunications Industry Association, the British trade association for the telecommunications industry and not to be confused with the American organisation of a similar name.

In general, European Union bodies spending public money should use CENELEC standards, where they exist, to specify their requirements to ensure fair access to that market by all European manufacturers.

CEN is another European body that works in partnership with CENELEC and ETSI. CEN's mission is, 'to promote voluntary technical harmonisation in Europe in conjunction with world-wide bodies and its partners in Europe'.

ETSI is the European Telecommunications Standards Institute

based in Sophia Antipolis in Southern France. It produces voluntary telecommunications standards in response to requests from its members, currently numbering 700 across 50 countries.

## 15.1.3  America

The United States of America contains many standards writing organisations.

### *ANSI (American National Standards Institute)*

ANSI publishes many standards including test methods and LANs. The TIA and EIA publish their standards under the auspices of ANSI so that they are ANSI/TIA/EIA standards. ANSI describes itself as the administrator and co-ordinator of the US private sector voluntary standardisation system since 1918. ANSI does not in itself develop American National Standards (ANS) but rather facilitates development by establishing consensus among qualified groups. ANSI accredits more than 175 entities to actually develop the standards; bodies such as the TIA for example. ANSI was a founding member of the ISO and is the sole US representative as it is with the IEC via the US National Committee (USNC). ANSI also accredits the US TAGs (Technical Advisory Groups) whose primary purpose is to develop and transmit, via ANSI, US positions on activities and ballots of the international technical committee.

### *ASTM (American Society for Testing and Materials)*

The ASTM generates many basic electrical tests that are referenced in other more specific tests for structured cabling.

### *BICSI (Building Industry Consulting Services International)*

BICSI is a not-for-profit trade organisation dedicated to promoting professional qualifications within the cabling industry. BICSI now operates in 70 countries with over 15000 members and offers the professional RCDD, Registered Cable Distribution Designer qualification. BICSI does not write standards but it provides industry feedback to those that do and is currently working with NECA, National Electrical Contractors Association, to produce a cabling installation standard to be known as ANSI/NECA/BICSI Commercial Building Telecommunications Cabling Installation Standard No. 300.

## EIA (Electrical Industry Alliance)

The EIA writes more specific component based electrical standards such as EIA-310, racks and cabinets. The EIA describes itself as providing a forum for industry to develop standards and publications in major technical areas: electronic components, consumer electronics, electronic information and telecommunications.

## FCC (Federal Communications Commission)

The FCC lays down many rules regarding the use of telecommunications equipment within the United States. Two sets of rules particularly apply to the use of structured cabling systems; FCC part 15, Electromagnetic Radiation issues and FCC part 68, Connection of Premises Equipment and Wiring to the Telecommunications Network.

## ICEA (Insulated Cable Engineers Association)

ICEA is made up of cable manufacturers and produces the Telecommunications Wire and Cable Standards — Technical Advisory Committee, (TWCS-TAC). The TIA may adopt or refer to these standards for cables, if ANSI approved, to avoid duplication of work on cable technology.

## IEEE (Institute of Electrical and Electronic Engineers)

The IEEE is a professional body whose main claim to fame in this arena is writing most of the LAN standards in use today. For example the ubiquitous Ethernet standard comes from the IEEE 802.3 committee. The IEEE is concerned with cable system performance, as it is the physical layer of the stack of network protocols.

The IEEE, with its 330000 members in 150 countries, produces more than thirty percent of the world's published literature in electrical engineering, computers and control technology.

The IEEE also publishes the National Electrical Safety Code (NESC) which covers basic provisions for the safeguarding of persons from the hazards arising from the installation, operation or maintenance of outside plant.

## NEMA (National Electrical Manufacturers Association)

NEMA produces cable specifications from its *High performance wire and cable section, premises wiring* section. The TIA may adopt some of the NEMA work for inclusion into its wider, system-based standard. NEMA standards start with 'WC' and the most appropriate is WC-66.1 *Category 6 and 7 one-hundred ohm shielded and unshielded twisted pair cable*. The NEMA cable standards are scheduled for ANSI approval and will probably be referenced by the TIA.

## NFPA (National Fire Protection Association)

The NFPA has spent the last one hundred years developing codes and standards concerning all area of fire safety. It now has 65000 members in 70 countries. There are currently more than 300 NFPA fire codes in use, such as:

- NFPA 1 — Fire Prevention Code.
- NFPA 54 — National Fuel Gas Code.
- NFPA 70 — National Electrical Code.
- NFPA 101 — Life Safety Code.

NFPA 70, with its National Electrical Code and various articles is the most prevalent code within the structured cabling industry.

## TIA (Telecommunications Industry Association)

The TIA is a trade association active since 1924. The TIA attempts to represent the American telecommunications industry along with its subsidiary, the MultiMedia Telecommunications Association (MMTA), and in conjunction with the Electronic Industries Alliance, the EIA. The TIA has five product-oriented divisions, 'premises equipment', 'network equipment', 'wireless communications', 'fibre optics' and 'satellite communications'. Each division prepares standards dealing with performance testing and compatibility.

The TIA maintains standards formulating groups. The most important one for structured cabling is TR 42, 'User premises telecommunications cabling infrastructure'. Another relevant group is TR41, 'User premises telecom requirements'.

Most of the major American standards relating to structured cabling are published under the auspices of ANSI/TIA/EIA. Figure 15.1 shows the organisation chart of the group known as TR42, and TR41. The TR42 group within the TIA structure is very influential for publishing Standards, Addenda, Revised Standards and Telecommunications Systems Bulletins (TSBs) involving commercial building cabling, e.g. TIA-568-A; also pathways and spaces, TIA-569-A and administration and labelling, TIA-606-A. TR 41 publishes the grounding and bonding standard TIA-607-A.

The TIA also maintains two fibre optic divisions, FO-2 and FO-6. FO-2 group develops physical-layer optical fiber system test procedures (OFSTP) TIA-526 series. The FO-6 group develops standard fiber optic test procedures (FOTP) TIA-455 series, informative test methods and fiber optic connector intermateability standards (FOCIS).

## Underwriters Laboratories Inc (UL)

UL is an independent, not-for-profit product safety testing and certification organisation. UL marks appear on about 15 billion items every year and 90000 new products are evaluated every year. UL develops tests such as in UL 910, the plenum fire rating test.

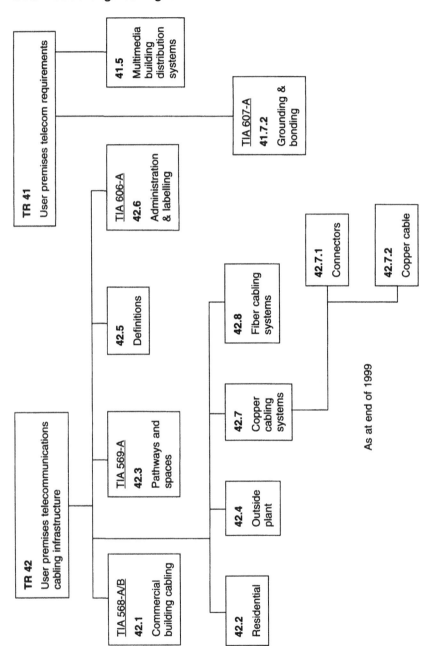

**Fig. 15.1** Organisation of TIA TR 42.

Note that tested products can be described as 'listed', 'classified' or 'recognised':

- A product can be *listed* after it has successfully passed a series of electrical and mechanical tests that simulate all likely hazards to be encountered by that product.
- A product can be *recognised* after it has been tested and passes for use as a component in a listed package.
- A product is *classified* if it is evaluated and passes tests for specific hazards or certain regulatory codes.

## 15.1.4   Canada

Standards development, certification and testing in Canada comes within the auspices of CSA International. Canadian cabling standards tend to 'mirror' American standards to a large extent, for example the ANSI/TIA/EIA 568-A specification has a version published in Canada known as CSA T529.

The Canadian Electrical Code part 1 (CEC pt 1) includes safety standards to be maintained when installing cables.

## 15.1.5   Australia and New Zealand

Standards Australia and Standards New Zealand work together to produce joint publications. For structured cabling the ISO 11801 standard has been amended to produce AS/NZS 3080:1996 *Telecommunications installations — integrated telecommunications cabling systems for commercial premises*. 'This standard reduces the choices available in Australia and New Zealand to recognise existing national directions and infrastructures, and also to avoid compatibility problems such as those likely to occur if more than one of the options allowable under the International Standard are implemented'.

Also used in Australia/NZ is HB29 *Telecommunications cabling handbook*, AS/NZS 3085.1 *Administration standard* and AS 3084, *Pathways and spaces*, based on TIA/EIA 569-A. The new IEC 61935 test standard will be adopted without local amendments.

### 15.1.6   International

## *ATM forum*

The ATM forum is an international non-profit organisation composed of mainly network equipment manufacturers with the aim of promoting the use of ATM (asynchronous transfer mode) networking technology by agreeing standards and common interface requirements. The ATM forum consists of a worldwide technical committee, three marketing committees and a user committee, through which ATM end users can participate.

The ATM forum was created in 1991 and now has over 600 members. Dozens of technical specifications have been published by the ATM forum covering issues such as control signalling, LAN emulation, network management and physical layer. It is the latter that stipulates electrical and optical performance requirements of the cabling system.

## *IEC (International Electrotechnical Commission)*

The IEC was founded in 1906 as a result of the 1904 International Electrical Congress held in St Louis. Fifty countries participate to prepare and publish international standards for all electrical, electronic and related technologies.

The IEC charter includes electronics, magnetics, electroacoustics, telecommunications and energy productions as well as associated disciplines such as terminology, symbols and measurements.

The IEC is one of the bodies recognised by the World Trade Organisation and entrusted by it for monitoring the national and regional organisations agreeing to use the IEC's international standards as the basis for national or regional standards as part of the WTO's technical barriers to trade agreement.

## *ISO (International Organisation for Standardisation)*

Formed in 1947 as a non-governmental organisation based in Geneva somewhat mysteriously calls itself ISO rather than IOS, concentrating on the Greek derived *isos* meaning 'equal.'

ISO standards cover all technical fields except electrical and electronic engineering, which is the responsibility of the IEC. A joint ISO/IEC technical committee called JTC 1 carries out the work in the field of Information Technology. The 'famous' ISO 11801 standard is actually entitled ISO/IEC 11801 and is prepared by Joint Technical Committee No. 1, Sub Committee No. 25, Working Group No. 3, or, ISO/IEC JTC1 SC25 WG3.

Some of ISO's other well known standards are the internationally recognised quality system ISO 9000, and the ISO 14000 Environmental Management system is also gaining wide acceptance. The seven base units of the Système International d'Unités such as metres and kilograms are covered by fourteen International Standards. To date there are over 12 000 International Standards representing more than 300 000 pages of information.

## ITU (International Telecommunications Union)

The ITU was formerly known as CCITT, and like many other standards bodies is located in Geneva, Switzerland. The ITU provides a forum within which governments and the private sector can coordinate global telecom networks and services. The ITU-T fulfils the purposes of the ITU relating to telecommunications standardisation by studying technical, operating and tariff questions and adopting Recommendations (note: not called 'standards') with a view to standardising telecommunications on a world-wide basis. An example of recent ITU-T work is the UIFN or universal international freephone number.

## 15.2   Cable system design

There are several standards that cover the design of a structured cabling system, e.g. ISO 11801. At the simplest level one could simply quote ISO 11801 as a requirement and be safe in the knowledge that a complete structured cabling system is specified therein. It is not the complete story however. ISO 11801 and its sister publications such as TIA 568 either refer to other standards or leave the user to pick out options for themselves, principally:

- What optical fibre type to use?
- What kind of optical connector to use?
- The use of screened or unscreened copper cable.
- Grade and type of copper or optical cable to use to the desk (horizontal).
- Grade and type of copper or optical cable to use in the backbone.
- Density of cabling and number of outlets.
- Cable containment systems.
- Fire safety performance of the cabling.
- Type and methods of testing.
- Quality and administration standard.
- EMC/EMI performance.
- Method of grounding and bonding.
- Limitations of LAN standards superimposed upon premises cabling standards.
- Legal obligations of directives and codes.
- Which standards apply in which territory?

It is not the intention of this book to act as a complete cable system design manual but instead to explain the technology, engineering and physics which lay behind the cabling standards and to explain which standards are appropriate and where.

## 15.2.1   Territories

In general, users should quote standards most relevant to the territory in which the cabling will be installed. If there is no national standard then international standards should be invoked. It is permissible to quote several standards and require the supplier to conform to the most severe requirements of all of them. This chapter deals with the following groups of standards:

- European Union.
- United States of America.
- Canada.
- Australia.

- New Zealand.
- International.

Users in countries not specifically listed should use international standards or request information and advice from their national standards body. A list of these is included in Appendix II.

## *Principal design standards*

| European Union | United States | Canada | Australia & New Zealand | Rest of the World |
|---|---|---|---|---|
| EN 50173 | TIA/EIA 568-A | CAN/CSA- T529 | AS/NZS 3080 | ISO 11801 |

*Notes*:

EN 50173, *Information technology, generic cabling systems*, August 1995

This first edition of EN 50173 was substantially changed in January 2000 by the addition of Amendment 1, which raises the values of class D cabling to take into account the requirements of gigabit Ethernet. In 2002 the 2nd edition will be published which will fully describe categories 5, 6 and 7 (classes D, E and F) cabling. EN 50173 1st edition references EN 50167, EN 50168 and EN 50169 as cable specifications. The 2nd edition will reference EN 50288 for cables, EN 50174 for cabling installation and quality assurance and EN 50xxx for testing methods and requirements.

TIA/EIA-568-A, *Commercial building telecommunications cabling standard*, October 1995

TIA/EIA-568-A has been substantially revised and added to as follows:

| Addendum 1 | Propagation delay and delay skew | Sep 1997 |
| Addendum 2 | Corrections and additions to TIA-568-A | Aug 1998 |

Addendum 3    Hybrid cables                              Dec 1998
Addendum 4    Patch cord qualification test              Aug 1999
Addendum 5    Additional transmission performance        Nov 1999
              specifications for 4-pair 100 $\Omega$
              category 5e cabling

The main standards from the TIA are supplemented from time to time by Telecommunications Systems Bulletins or TSBs:

TSB67    Transmission performance specifications for    Oct 1995
         field testing of twisted pair cabling
         systems
TSB72    Centralized optical fiber cabling guidelines    Oct 1995
TSB75    Additional horizontal cabling practices for     Aug 1996
         open offices
TSB95    Additional transmission performance             Aug 1999
         guidelines for 100 $\Omega$ 4-pair category
         5 cabling

There is also a relevant interim standard called:

TIA-IS 729            Additional requirements for 100 $\Omega$    Mar 1999
                      screened twisted pair cabling

TIA 568-A calls upon:

ANSI/EIA/TIA-569    Commercial building standard for telecommuni-
                    cations pathways and spaces.
ANSI/EIA/TIA-570    Residential and light commercial telecommuni-
                    cations wiring standard.
ANSI/EIA/TIA-606    Administration standard for the telecommunica-
                    tions infrastructure of commercial buildings.
ANSI/EIA/TIA-607    Commercial    building    grounding/bonding
                    requirements.

In 2000, all of the Addenda and TSBs related to TIA 568-A will be rounded up in a brand new standard called EIA/TIA 568-B. 568-B will be published in three parts:

- TIA-568-B.1   Commercial cabling standard, master document.
- TIA-568-B.2   Twisted pair media.
- TIA-568-B.3   Optical fiber cabling standard.

Some of the changes in the new 568-B standard will include the addition of screened (shielded) cabling, 50/125 optical fibre and new optical connectors. 568-B will not include category 6; that will be published first as an interim standard that is currently known as Proposal No. PN-3727.

ISO/IEC 11801, *Information technology — generic cabling for customer premises*, July 1995

The 1995 publication is known as the first edition. It was amended in 1999 with amendments one and two to make corrections but principally to raise the specification of class D cabling to make it gigabit Ethernet compliant. The second edition will be published in 2001 to incorporate category 6 and 7 cabling and to seek as much common wording as possible with American standards.

# 15.3   Component standards

The design standards will call upon individual component standards that may or may not be part of the design standard.

| Territory | European Union | United States | Rest of the World |
|---|---|---|---|
| System design standard | EN 50173 | TIA/EIA 568* | ISO 11801 |
| Component standard | *Copper cable* EN 50288 *Copper connector* IEC 60603-7 *Optical fibre* EN 188 000 *Optical cable* EN 187 000 *Optical connectors* EN 186 000 | To be specified in TIA/EIA 568B:2000 | *Copper cable* IEC 61156 *Copper connector* IEC 60603-7 *Optical fibre* IEC 60793 *Optical cable* IEC 60794 *Optical connectors* IEC 60874 |

*Note*: TSB 36 (category 5 cables) and TSB 40A (additional transmission specifications for unshielded twisted pair connecting hardware) are still sometimes invoked but they have been absorbed into TIA 568-A which in turn will be replaced by TIA 568-B.

*Note*: Copper cables
European cables are also defined in the following publications:

EN 50167    Sectional specifications for horizontal floor wiring cables with a common overall screen for use in digital communications.

EN 50168    Sectional specifications for work-area wiring cables with a common overall screen for use in digital communications.

EN 50169    Sectional specifications for backbone cables, riser and campus with a common overall screen for use in digital communications.

The above three standards are being replaced by the following:

EN 50288-2-1    100MHz screened, horizontal and backbone.
EN 50288-2-2    100MHz screened, patch.
EN 50288-3-1    100MHz unscreened, horizontal and backbone.
EN 50288-3-2    100MHz unscreened, patch.
EN 50288-4-1    600MHz screened, horizontal and backbone.
EN 50288-4-2    600MHz screened, patch.
EN 50288-5-1    200MHz screened, horizontal and backbone.
EN 50288-5-2    200MHz screened, patch.
EN 50288-6-1    200MHz unscreened, horizontal and backbone.
EN 50288-6-2    200MHz unscreened, patch.

Cables are also defined in the following American standards:

NEMA WC-63.1    Performance standards for twisted pair premise voice and data communications cable.

NEMA WC-63.2    Performance standards for coaxial communications cable.

NEMA WC-66.1    Performance standards for category 6, category 7 $100\,\Omega$ shielded and unshielded twisted pair cables.

| ICEA S-80-576 | Communications wire and cable for wiring of premises. |
| ICEA S-89-648 | Aerial service wire. |
| ICEA S-90-661 | Individually unshielded twisted pair indoor cables. |
| ICEA S-100-685 | Station wire for indoor/outdoor use. |
| ICEA S-101-699 | Category 3 station wire and inside wiring cables up to 600 pairs. |
| ICEA S-102-700 | Category 5, 4-pair, indoor UTP wiring standard. |
| ICEA S-103-701 | AR & M riser cable. |

*Note*: proposed IEC connector standards structure

| 60603-7 | Unscreened | 3 MHz |
| 60603-7-1 | Screened | 3 MHz |
| 60603-7-2 | Unscreened | 100 MHz cat5e |
| 60603-7-3 | Screened | 100 MHz cat5e |
| 60603-7-4 | Unscreened | 250 MHz cat6 |
| 60603-7-5 | Screened | 250 MHz cat6 |
| 60603-7-6 | Unscreened | 600 MHz cat7 |
| 60603-7-7 | Screened | 600 MHz cat7 |

*Note*: Optical cables
   Some American optical cable specifications are:

| ICEA S-87-640 | Optical fiber outside plant cable. |
| ICEA S-83-596 | Optical fiber indoor/outdoor cable. |

The structure of IEC 60794 is:

| IEC 60794-1-2 | Testing |
| IEC 60794-2 | Internal cables. |
| IEC 60794-2-10 | Simplex and duplex cables. |
| IEC 60794-2-20 | Multi-fibre cables. |
| IEC 60794-2-30 | Ribbon cords. |
| IEC 60794-3 | External cables |
| IEC 60794-3-10 | Duct or buried cables. |
| IEC 60794-3-20 | Aerial cables. |
| IEC 60794-3-30 | Underwater cables. |
| IEC 60794-4 | Cables along electrical overhead lines. |

## 15.4   Fire performance standards

In the US there are fire performance requirements based on where-abouts in a building a cable has been installed. The most demand-ing is the plenum zone, i.e. the space, usually inside a false ceiling, where environmental air is being circulated. The second most demanding area is called the riser and apart from that cables are specified as general purpose. The cables have to be specifically marked according to their designation, e.g. OFNP marking means optical fiber non-conductive plenum. The full list of optical cable types and their respective test methods are summarised in Table 15.1.

Although Plenum cable has a very low level of flammability, and is the most severe cable test around, it suffers from two disadvantages; plenum grade material is very expensive and secondly the material is fluorinated, typically being PTFE (poly tetra fluoro ethylene), i.e. Teflon. Plenum cable therefore has no chance of passing a zero halogen test requirement such as IEC 60754.

In Europe, as yet, there are no flammability requirements for cables, but this will change with the introduction of the construction prod-ucts directive in 2001. The relevant IEC standards for fire performance are:

IEC 60332-1      Flammability of a single vertical cable.
IEC 60332-3-c    Flammability of a bunch of vertical cables.
IEC 61034        Smoke density and evolution.

| Table 15.1 American optical cable marking | | | |
|---|---|---|---|
| Cable title | Marking | Test method | |
| Conductive optical fiber cable | OFC | General purpose UL 1581 | |
| Non conductive optical fiber cable | OFN | General purpose UL 1581 | |
| Conductive riser | OFCR | Riser | UL 1666 |
| Non conductive riser | OFNR | Riser | UL 1666 |
| Conductive plenum | OFCP | Plenum | UL 910 |
| Non conductive plenum | OFNP | Plenum | UL 910 |

IEC 60754-1      Halogen gas emission.
IEC 60754-2      Smoke corrosivity.

Also sometimes invoked are:

Smoke density: ASTM E662, NES 713.

It must be remembered when specifying cable fire performance that some of the standards are mutually exclusive, for example one cannot specify UL 910 *and* IEC 60754-1.

## 15.5   EMC/EMI standards (including the 'screened versus unscreened' debate)

| Territory | European Union | United States | Rest of the World |
|---|---|---|---|
| System design standard | EN 50173 | TIA/EIA 568 | ISO 11801 |
| EMC/EMI standard | EN 50081 EN 50082 EN 55022 EN 55105 | FCC part 15 | EN, FCC or local standard |

EMC or electromagnetic compatibility is viewed as an electromagnetic emissions requirement whereas EMI, electromagnetic immunity, is a measure of the device or system's capability at rejecting outside electromagnetic interference. The term EMC is often used to cover both radiated and received electrical noise.

The UK Electromagnetic Compatibility Regulations (Statutory Instrument 1992/2372) enacted the European EMC Directive in UK law and as from 1st January 1996 information technology equipment must meet the relevant European EMC standard and carry a 'CE' mark indicating compliance.

The CE mark shows that a product meets all required European specifications, not just EMC. All active electronic equipment must carry the CE mark to demonstrate compliance. The CE mark however cannot be put onto passive equipment such as cabling and associated hardware. What kind of signal the cabling carries and the length and disposition of that cabling can change from one day to the next. Any EMC test done on structured cabling, for example a typical cabling layout running 100baseT Ethernet, is of general interest but it only proves that particular combination of manufacturer's equipment on that particular layout of cabling.

The philosophy must be that a well-balanced and engineered cabling system must not degrade the EMC performance of any active equipment connected to it. 'CE' conformant LAN equipment must still present a completely conformant IT system when plugged into the cabling system. A cabling system cannot rectify a badly balanced electrical signal and will radiate under such conditions.

EN 50081 is a general emission standard and EN 50082 is a general immunity standard. EN 55022 relates specifically to the electromagnetic emissions of information technology equipment. EN 55105 (formerly prEN 55024) relates specifically to the immunity requirements of information technology equipment, excluding telecommunications terminal equipment.

There are other tests specified from time to time but they are usually component tests and the tests listed above should take precedence as they relate to complete systems. Other such tests include:

- EN 61000 (IEC 61000) electromagnetic compatibility, environment.
- IEC 801 Electrostatic discharge and electrical fast transient immunity.

The questions arise; radiating how much electromagnetic energy and at what frequency and immune to how much electromagnetic energy and across which frequency band?

Both FCC and EN standards address the issue of residential and industrial environment. EMC emissions are tighter for the residential environment than the industrial because it is considered that equipment will be located closer together in the residential environment.

Class A is the non-residential environment and is around 50 mV/m maximum radiated field strength measured at 30 m away and across the frequency band of interest (up to 216 MHz for FCC pt 15 and 230 MHz for EN 55022). Class B is the more demanding residential requirement of less than 150 mV/m but this time measured at just 3 m range.

Both the LAN equipment and cabling infrastructure should meet EN55022 and FCC part 15 class B requirements when CE/FCC compliant transmission equipment is being used.

The cabling system should be able to withstand an impinging electromagnetic field of up to 3 V/m across a band of dc to 1000 MHz without incurring an additional bit error rate beyond that expected for any particular LAN protocol across all its operating speeds. For example, an ATM 155 Mb/s signal sent with zero errors should not arrive with any more than one in ten to the power of ten errors when subject to the above interfering external signal.

How much 'real life interference' does three volts per metre represent? The following equation 15.1 approximates the field strength in V/m from a radiating source when the distance and power output of the source are known:

$$\text{field strength} \approx \frac{\sqrt{30 \times W}}{d} \qquad [15.1]$$

where $W$ = output power of source in watts
$d$ = distance from source in metres

Another approximation sometimes used is $E = \frac{7 \times \sqrt{W}}{d} V/m$. The nature of the antenna, e.g. half-wave dipole, will also make a difference.[2] We can see that a mobile telephone, from a few metres, and an airport radar (600 kW) 1000 m away are capable of generating around 4 V/m field strength.

Tests done by 3P Laboratories in Denmark[1], whereby screened and unscreened cables were subject to 3 V/m fields showed that peaks up to 35 mV were induced into unscreened cables (UTP) but only a few millivolts were induced into a screened cable. Is this significant? For robust protocols, such as 10baseT, probably not. But for the new multilevel coding schemes, like the 5 level PAM of 1000baseT, it may be.

Roughly speaking, for a 2V peak-to-peak signal at the transmitter and using a five level code, the difference between any two adjacent groups of code, i.e. voltage levels, is going to be about half a volt. Given the attenuation of the cabling channel, what length of cable will it take to reduce the half a volt, or 500 mV, to the level where it is of a similar magnitude to the 30 millivolts or so that appear to make it into the cable with a 3 V/m field impinging onto a UTP cabling system?

To reduce 500 mV to 30 mV requires 24 dB of attenuation, and this just happens to be the channel attenuation of a 100 m class D channel at 100 MHz. To reinforce the idea that 30 mV is of an important magnitude we can see that the ATM 155 Mb/s standard, AF-PHY-0015-00,0 paragraph 5.3.1, calls for no more than 20 mV noise to appear on the cable and the 1000baseT standard IEEE 802.3ab, calls for no more than 40 mV (40.7.5.1).

What happens if the field increases to 10 V/m? We can presume the induced voltage in the cable will increase to around 100 mV. It takes 14 dB of attenuation to reduce 500 mV to 100 mV. 14 dB, at 0.22 dB/m at 100 MHz for the cable attenuation, means that the signal voltage differential has been reduced to the noise level at only 34 m.

The practical effect of occasional noise of this magnitude would not be significant. A burst of noise will be masked by the error correcting codes of the LAN protocol. If it happened frequently then the user would notice longer and longer response times as data blocks were repeatedly retransmitted. But eventually the network would cease to function.

How will the users know if he is being subject to a large interfering electromagnetic field? The practical effects, as we have said, will be long response times, system crashes, and for mainframe systems a mysterious logging on and off of terminals. But the only way to prove the existence of interference is to employ an EMC specialist who will set up antennas within the workplace and record the ambient electromagnetic noise environment across a suitable spectrum. The test should ideally be run across a full, normal, seven day working cycle so that any level of interference coincident with some obvious event, such as a machine starting, can be seen.

We can thus draw some conclusions from these hypotheses:

- Users running their system in an environment of 3 V/m or less ambient field with only occasional transgressions above that, should be able to use a well engineered and installed unscreened cable system at high LAN speeds up to and beyond 1000 Mb/s.
- Users operating in an environment of sustained interfering field of 3–10 V/m and wish to run high-speed gigabit LANs and with distances in excess of around 40 m should consider a screened cable system.
- Users operating in an EMC environment above 10 V/m should consider using optical fibre.

Other practical issues to be taken into account are to install the cabling as far away from sources of interference as possible, e.g. fluorescent lamps, power cables, lift motors, switching gear, mobile 'phone users and any other radio or television transmitting equipment. See EN 50174 for more details of minimum spacing of cables from sources of potential interference.

# 15.6   Test standards

| Territory | European Union | United States | Rest of the World |
|---|---|---|---|
| System design standard | EN 50173 | TIA/EIA 568 | ISO 11801 |
| Testing and acceptance standard | EN 50174 EN 50xxx* | TSB 67 TSB 95 568-A Addendum 5 | ISO 14763 IEC 61935 IEC 61280 IEC 60793 |

*Note: EN 50xxx will reference IEC standards such as IEC 61935.

Table 15.2 gives the parameters to be measured according to the different standards. It should be noted that for 'old' category 5 installations (or below) then the test parameters specified in the TSB 67 column should be used. To test existing installed category 5 cabling

Table 15.2 Test requirements from different standards

| | ANSI/TIA/EIA 568A | | | TIA/EIA 568B | IEC 61935 | ISO 11801 2nd edn |
|---|---|---|---|---|---|---|
| | TSB 67 | TSB 95 | Addendum 5 | | | |
| Wire map | x | x | x | x | x | x |
| Length | x | x | x | x | | x |
| Attenuation | x | x | x | x | x | x |
| NEXT pair to pair | x | x | x | x | x | x |
| NEXT powersum | | | x | x | x | |
| ELFEXT pair to pair | | x | x | x | x | x |
| ELFEXT powersum | | x | x | x | x | |
| Return loss | | x | x | x | x | |
| Propagation delay | | x | x | x | x | x |
| Delay skew | | x | x | x | x | x |
| DC loop resistance | | | | | x | |
| Shield continuity | | | | | | x |

to see if it is gigabit Ethernet compatible then the standard called 'TSB 95' should be used. For enhanced category 5, category 6 or 7 then the parameters listed in TIA 568-A Addendum 5, TIA 568-B or IEC 61935 should be used.

ISO 11801 2nd edition lists tests under the titles of 'acceptance' and 'reference', Table 15.3. The installer should do acceptance and compliance but the reference tests (see prEN 50289) must be done by the equipment manufacturer and test reports should be available to the end user and installer for inspection upon request. ISO 11801 2nd edition also references IEC 61935.

The only acceptance test for optical fibre is an end-to-end link optical attenuation test (sometimes return loss as well). But note that this must be done at every wavelength of intended operation, such as 850 nm and 1300 nm and preferably in both directions. The standards are not clear about testing from both directions but in the

Table 15.3 Acceptance and reference standards from ISO11801:2000

|  | Acceptance and compliance | Reference |
|---|---|---|
| Propagation delay | x | x |
| Skew | x | x |
| DC loop resistance |  | x |
| NEXT | x | x |
| ELFEXT | x | x |
| Attenuation | x | x |
| Return loss | x | x |
| Longitudinal conversion loss |  | x |
| Transfer impedance |  | x |
| Coupling attenuation |  | x |
| Shield dc resistance |  | x |
| Conductor map | x | x |
| Length | x | x |
| Continuity of conductors and shields | x | x |

Adapted from draft ISO/IEC 11801:2000, March 1999, Table A.1.

author's opinion it is highly recommended to get a true indication of the link performance.

The chapters dealing with optical and copper cable technology give more details of the methods involved in cable testing and recording of results.

# 15.7  Earthing, grounding and bonding

| Territory | European Union | United States | Rest of the World |
|---|---|---|---|
| System design standard | EN 50173 | TIA/EIA 568 | ISO 11801 |
| Earthing, grounding and bonding standard | EN 50174<br>EN 50303<br>EN 60950 | TIA 607 | IEC 60364<br>IEC 61024<br>ITU-T K31 |

All metallic cable containment systems and cable rack must be continuously grounded and bonded according to international or local standards, such as:

PrEN50303    Application of equipotential bonding and earthing at premises with information technology equipment.

PrEN50174-2    Information technology, cabling installation, part 2, installation, planning and practices inside buildings.

TIA/EIA-607    Commercial building grounding and bonding requirements for telecommunications.

All screened cables must be installed and grounded according to the manufacturer's instructions. The end-user is responsible for supplying a low impedance-to-earth grounding point at all relevant locations. The installer must identify where these locations are and bring to the attention of the end-user any point where such a grounding point is not available at any location that affects the safe and efficient use of the IT equipment.

# 15.8  Cable containment and administration

| Territory | European Union | United States | Rest of the World |
|---|---|---|---|
| System design standard | EN 50173 | TIA/EIA 568 | ISO 11801 |
| Cable containment and administration | EN 50174-2 | TIA 606 TIA 569 | ISO 14763-2 |

The cable administration and containment system must ensure that all cabling requirements, e.g. bend radii, imposed by the cable manufacturer, and any relevant international and local standards are adhered to, such as:

PrEN50174-2    Information technology, cabling installation, part 2, installation, planning and practices inside buildings.

ISO 14763-2    Information technology, implementation and operation of customer premises cabling, part 2, planning and installation.

TIA/EIA-606    Administration standard for the telecommunications infrastructure of commercial buildings.

## 15.9    Local area network standards

Local area networks (LANs) need a communications medium upon which to run. It may be wireless (IEEE 802.11) but in most cases it will be on copper or optical cable. The writers of LAN standards and cabling standards must work hand-in-hand; structured cabling only exists to carry communications traffic and LANs would be pointless without cabling. Although the structured cabling standards obviously address the cabling in detail this does not prevent the LAN standards from detailing exactly what electrical and optical performance is expected of the cabling system to ensure reliable operation of the LAN. Usually the LAN and cabling standards are mostly coincident but it should be noted that optical LANs cannot run over all the distances allowed in the structured cabling standards. Note also that IEEE 802.3ae, ten gigabit Ethernet, is expected in 2002. It will most likely be in the form of:

10000base-SX 850 nm.
10000base-LX 1300 nm.
10000base-EX 1550 nm.

The optical fibre performance requirement for multimode fibre to allow 10 000 Mb/s operation is yet to be finalised.

Nearly all LAN standards originate from the IEEE 802 group, the ATM Forum or other ANSI accredited groups. Figure 15.2 shows the construction of the IEEE 802 committee. Table 15.4 gives all the LAN standards that specify a physical layer, i.e. a cable performance requirement. Note that IEEE 802 standards are adopted as ISO/IEC

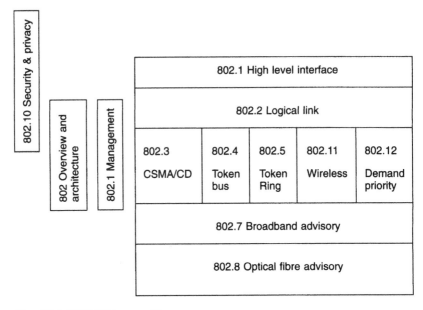

**Fig. 15.2** IEEE 802 committee summary.

standards and have an extra '8' added to the beginning, so IEEE 802.3, for example, becomes ISO/IEC 8802.3.

The 802 committees are:

- 802.1 High level interface. Defines the relationship between IEEE 802 standards and higher levels of the OSI reference model.
- 802.2 Logical link control.
- 802.3 Ethernet CSMA/CD. The physical layer standard defining Carrier Sense, Multiple Access and Collision Detection.
- 802.4 Token bus. A token passing access method with a bus topology. Also known as Manufacturing Automation Protocol (MAP) and ARCnet.
- 802.5 Token ring. A token passing access method on a ring topology. Also known as HSTR, High Speed Token Ring, for the later 100 and 1000 Mb/s proposals.
- 802.6 Metropolitan area network. A MAN topology known as distributed queue dual bus.
- 802.7 Broadband advisory group. This group gives advice to all other subcommittees on broadband networking techniques.

Table 15.4  LAN applications with specified physical media requirements

Class A (100 kHz)

| Application | Specification | Date | Additional name |
|---|---|---|---|
| PBX | National requirements | | |
| X21 | ITU-T Rec. X.21 | 1994 | |
| V11 | ITU-T Rec. X.21 | 1994 | |

Class B 1 MHz

| Application | Specification | Date | Additional Name |
|---|---|---|---|
| S-Bus | ITU-T Rec. I.430 | 1993 | ISDN Basic Access |
| $S_1/S_2$ | ITU-T Rec. I.431 | 1993 | ISDN Primary Access |

Class C 16 MHz

| Application | Specification | Date | Additional Name |
|---|---|---|---|
| CSMA/CD 10BaseT | ISO/IEC 8802-3 | 1996 | |
| Token Ring 4 Mb/s | ISO/IEC 8802-5 | 1995 | |
| ISLAN | ISO/IEC 8802-9 | 1996 | Integrated Services LAN |
| Demand Priority | ISO/IEC DIS 8802-12 | 1997 | VGAnyLAN |
| ATM LAN 25.6 Mb/s | ATM Forum af-phy-0040.000 | 1995 | ATM-25 Cat 3 |
| ATM LAN 51.84 Mb/s | ATM Forum af-phy-0018.000 | 1994 | ATM-52 Cat 3 |
| ATM LAN 155.52 Mb/s | ATM Forum af-phy-0047.000 | 1995 | ATM-155 Cat 3 |
| CSMA/CD | ISO/IEC 8802-3 DAM21 | 1997 | Fast Ethernet |
| 100BaseT4 | IEEE 802.3u | 1995 | |
| CSMA/CD | ISO/IEC 8802-3 PDAM25 | 1997 | Fast Ethernet |
| 100BaseT2 | IEEE 802.3y | 1997 | |
| ISLAN 16-T | ISO/IEC 8802-9 DAM1 | 1997 | Isochronous Ethernet |
| | IEEE 802.9a | 1995 | |

Class D 100 MHz        1995 Standard for Class D/Cat5

| Application | Specification | Date | Additional Name |
|---|---|---|---|
| Token Ring 16 Mb/s | ISO/IEC 8802-5 | 1995 | |
| TP-PMD | ISO/IEC DIS 9314-10 | 1997 | Twisted-Pair Physical Medium Dependent |
| CSMA/CD | ISO/IEC 8802-3 DAM21 | 1997 | Fast Ethernet |
| 100BaseTX | IEEE 802.3u | 1995 | |
| ATM LAN 155.52 Mb/s | ATM Forum af-phy-0015.000 | 1994 | ATM-155 Cat 5 |

Table 15.4 *Continued*

| Class D 100MHz | 1999 Standard for Class D/Cat5e | | |
|---|---|---|---|
| Application | Specification | Date | Additional name |
| CSMA/CD 1000BaseT | IEEE 802.3ab | 1999 | Gigabit Ethernet |

Optical link

| Application | Specification | Date | Additional name |
|---|---|---|---|
| CSMA/CD FOIRL | ISO/IEC 8802-3 | 1996 | Fibre Optic Inter-Repeater Link |
| CSMA/CD 10BaseF | ISO/IEC 8802-3 | 1996 | Fibre Optic CSMA/CD |
| FDDI PMD | ISO/IEC 9314-3 | 1990 | Fibre Distributed Data Interface Physical Medium dependent |
| FDDI SMF-PMD | ISO/IEC 9314-4 | 1997 | Fibre Distributed Data Interface Single Mode Fibre PMD |
| FDDI LCF-PMD | ISO/IEC 9314-9 | 1997 | Fibre Distributed Data Interface Low Cost Fibre PMD |
| HIPPI | ISO/IEC 11518-1 | 1995 | High Performance Parallel Interface |
| ATM LAN 155.52Mb/s | ATM Forum af-phy-0062.000 | 1996 | ATM-155 multimode optical fibre |
| ATM LAN 622.08Mb/s | ATM Forum af-phy-0046.000 | 1996 | ATM-622 multimode optical fibre |
| CSMA/CD 1000BaseSX | IEEE 802.3z | 1998 | Gigabit Ethernet |
| CSMA/CD 1000BaseLX | IEEE 802.3z | 1998 | Gigabit Ethernet |
| Token Ring | ISO/IEC 8802-5 DAM2 | 1997 | Fibre Optic Station Attachment |
| Fibre Channel | ISO/IEC CD 14165-1 | 1997 | Fibre Channel Physical Layer |

Adapted from Table G of Annex G, ISO 11801, 2nd edn Draft, 1999.

- 802.8  Optical fibre advisory group. This group gives advice to all other subcommittees on optical fibre networking techniques.
- 802.9  Integrated services LAN. This committee works on the integration of voice, data and video traffic over other 802 LANs.

- 802.10   LAN security. This group works on standard security techniques for LANs using authentication and encryption techniques.
- 802.11   Wireless. Responsible for the standardisation of communications over spread-spectrum radio, narrowband radio, infrared light and utility power lines. 802.11a standard uses a 5 GHz band based on the ETSI HIPERLAN with bit rates of 6, 12 and 24 Mb/s. 802.11b uses the 2.45 GHz band with bit rates of 5.5–11 Mb/s.
- 802.12   Demand priority. The 100 Mb/s LAN standard known as 100VGAnyLAN which uses the Demand Priority access method.
- 802.14   Cable TV broadband. The IEEE committee which addresses digital communications services over cable television distribution networks.
- 802.15   Wireless personal area networks. Defines a wireless personal area network (PAN) working over about 10 m. Two industry groups are developing this technology and products; one is called Bluetooth and the other is called HomeRF.
- 802.16   Broadband wireless access. This group is developing a broadband air interface that will offer an alternative to the wired 'last mile'. It will work in the 24, 28, 31 or 40 GHz bands with ranges of a few kilometres. In the United States it is also referred to as local multipoint distribution service (LMDS).

# 15.10   Building automation standards

Structured cabling is generally associated with Local Area Networks, but this is not the only communications application within a building that requires cabling. We may also see:

- Telephony, PABX, PBX.
- Security video/CCTV.
- Public address systems.
- Fire/smoke detection.
- Fire alarms.
- Movement/security/PIR sensors.

- Access controls.
- HVAC (heating, ventilation and air conditioning).

All of these applications are capable of running on a structured cabling although most of them still use their own dedicated cabling media, from simple twisted pairs for telephony to coaxial cable for CCTV. Some of the more sophisticated control systems use 10baseT as the communications medium so their output and distribution becomes indistinguishable from any other kind of LAN traffic.

There is a good deal of conservatism in the building industry when it comes to the area generically known as 'intelligent buildings' or 'building automation systems'. Security and reliability are often stated as a problem factor due to all services running over just one cabling system. Topographically most LANs are now configured to operate in the star-wired format of structured cabling but most building automation systems are polled devices, using very low speeds, with cable daisy-chained from one device to another.

Another factor is the issue of definition of just what exactly does 'intelligent buildings' mean and where are they? We could for example have residential or home automation schemes, SoHo or small office/home office systems, commercial building automation and industrial or factory automation. All or any of these may use a basic LAN standard such as 10baseT or adopt their own unique standards and they may or may not use a standardised structured cabling scheme. Wireless schemes such as 802.15 wireless personal area networks (WPANs) are also a possibility. There are compelling economic reasons for centralising building management systems and utilising a common set of cables and so more and more integration of building automation and the structured cabling system is inevitable.

Some standards in use relevant to this area are: IEEE 1394 'Firewire', X10, LonWorks, CEBus, HomeRF, HomePNA, Universal serial bus, SCSI, EHS, BatiBus, Profibus, CANopen, InterBus, DeviceNet, EiBus, Euridis.

Some of the wider known standards bodies are working on:

- CENELEC prEN 50090 Home and Building Electronic Systems.
- CENELEC EN 60730 Automatic electric controls for household and similar applications.

- CENELEC EN 50170 General purpose field communication system.
- CEN CEN/TC247 Controls for mechanical building services.
- ISO/IEC CD 10192-1 Information technology — home electronic systems.
- Video electronics standards association (VESA) home networks committee.
- IEEE 802.15   Wireless personal access network.
- TIA 570-A     Residential telecommunications cabling standard.
- TIA TR-41.5   Multimedia building distribution systems.
    - SP-3771 Multimedia premises reference architecture.
    - PN-4407 Residential gateway.
- TIA TR-42  Building automation system.
- ISO 15018 Integrated cabling for all services other than mains power in homes, SoHo and buildings.

# References

1  Bech E, 'Proposed cabling set-up for electromagnetic characterisation of cabling and EMC measurements on LAN systems', *Communications Cabling EC97*, A.L. Harmer (Ed), IOS Press 1997.
2  'Interference Levels in Aircraft at Radio Frequencies used by Portable Telephones', Report no. 9/40: 23-90-02. Civil Aviation Authority, England, May 2000.

# Appendix I: List of some relevant standards

Note that in some cases only the generic title is given, e.g. IEC 60794. It must be remembered that there are many sub-headings in many of these standards. The same standard can appear in different guises, e.g. when IEEE standards are adopted as ISO standards. Note that not all standards quoted here had been fully published at time of printing, and the standards issuing body should be checked for the latest status. A full listing can be obtained by visiting the on-line catalogues of the standards publishers. Addresses are given in appendix II.

| | |
|---|---|
| af-phy-0015.000 | ATM forum. Physical medium dependent interface specification for 155 Mbs over twisted pair cable. |
| af-phy-0018.000 | ATM forum. Mid range physical layer specification for category 3 UTP. |
| af-phy-0040.000 | ATM forum. Physical interface specification for 25.6 Mbs over twisted pair. |
| af-phy-0046.000 | ATM forum. 622.08 Mbs physical layer. |
| af-phy-0047.000 | ATM forum. 155.52 Mbs physical layer specification for category 3 UTP. |
| af-phy-0053.000 | ATM forum. 120 $\Omega$ addendum for 155 Mbs over twisted pair. |
| af-phy-0062.000 | ATM forum. 155 Mbs over MMF short wave length lasers. |

| | |
|---|---|
| af-phy-0079.000 | ATM forum. 155 Mbs over plastic optical fibre. |
| af-phy-0079.001 | ATM forum. 155 Mbs over hard clad polymer fibre. |
| af-phy-0110.000 | ATM forum. Physical layer high density glass optical fibre annex. |
| af-phy-0133.000 | ATM forum. 2.4 Gbs physical layer specification. |
| ANSI X3T9.3/91-005 | HIPPI physical layer. |
| BS 6701 | Code of practice for installation of apparatus intended for connection to certain telecommunications systems. |
| BS 7718 | Code of practice for fibre optic cabling. |
| DISC PD1002 | A guide to cabling in private telecommunications systems (TIA UK). |
| EN 28877 | (ISO/IEC 8877) Information technology. Telecommunications and information exchange between systems. Interface connector and contact assignments for ISDN Basic Access interface located at reference points S and T. |
| EN 50081 | Electromagnetic compatibility — generic emission standard: part 1: residential, commercial and light industrial; part 2: industrial environment. |
| EN 50082 | Electromagnetic compatibility — generic immunity standard: part 1: residential, commercial and light industrial; part 2: industrial environment. |
| EN 50085 | Cable trunking systems and cable ducting systems for electrical installations. |
| EN 50086 | Conduit systems for electrical installations. |
| EN 50098-1 | Customer premises cabling for information technology part 1: ISDN basic access. |
| EN 50130 | Alarm systems — part 4: electromagnetic compatibility — product family standard: Immunity requirements for components |

|            | of fire, intruder and social alarm systems. |
|------------|---------------------------------------------|
| EN 50167   | Sectional specifications for horizontal floor wiring cables with a common overall screen for use in digital communications. |
| EN 50168   | Sectional specifications for work-area wiring cables with a common overall screen for use in digital communications. |
| EN 50169   | Sectional specifications for backbone cables, riser and campus with a common overall screen for use in digital communications. |
| EN 50173   | Information technology, generic cabling systems, August 1995 with amendment 1 published in 2000. |
| EN 50174   | Information technology, cabling installation. |
| EN 50288   | Multi-element metallic cables used in analogue and digital communications and control. |
| EN 50289   | Communication cables — specification for test methods. |
| EN 50310   | Application of equipotential bonding and earthing in buildings with information technology equipment. |
| EN 55013   | Limits and methods of measurement of radio disturbance characteristics of broadcast receivers and associated equipment. |
| EN 55020   | Electromagnetic immunity of broadcast receivers and associated equipment. |
| EN 55024   | Information technology equipment, immunity characteristics. Limits and methods of measurement. |
| EN 55103   | Electromagnetic compatibility — product family standard for audio, video, audio-visual and entertainment lighting control system for professional use. |

| EN 60825 | Safety of laser products (IEC 60825). |
| EN 60950 | Safety of information technology equipment (IEC 60950). |
| EN 61000 | (IEC 61000) Electromagnetic compatibility environment (contains over 17 parts). |
| ETS 300 253 | Equipment engineering — earthing and bonding of telecommunications equipment in telecommunications centres. |
| ICEA S-80-576 | Communications wire and cable for wiring of premises. |
| ICEA S-83-596 | Optical fiber indoor/outdoor cable. |
| ICEA S-87-640 | Optical fiber outside plant cable. |
| ICEA S-89-648 | Aerial service wire. |
| ICEA S-90-661 | Individually unshielded twisted pair indoor cables. |
| ICEA S-100-685 | Station wire for indoor/outdoor use. |
| ICEA S-101-699 | Category 3 station wire and inside wiring cables up to 600 pairs. |
| ICEA S-102-700 | Category 5, 4-pair, indoor UTP wiring standard. |
| ICEA S-103-701 | AR & M riser cable. |
| IEEE 802.3ab | Physical layer specification for 1000 Mb/s operation on four pairs of category 5 or better balanced twisted pair cable (1000baseT), July 1999. |
| IEEE 802.3z | Media access control (MAC) parameters, physical layer, repeater and management parameters for 1000 mb/s operation, June 1998. |
| ISO/IEC 11801 | Information technology — generic cabling for customer premises, 1995, with addendum 1 and 2 published in 1999. |
| ISO 15018 | Integrated cabling for all services other than mains power in homes, SoHo and buildings. |
| IEC 60332-1 | Flammability of a single vertical cable. |
| IEC 60332-3-c | Flammability of a bunch of vertical cables. |

| | |
|---|---|
| IEC 60364 | Electrical installations of building — Part 5: chapter 548: Earthing arrangements and equipotential bonding for I.T. systems. |
| IEC 60603-7 | Amendment 1. Detail specification for connectors 8 way. Test methods and related requirements for use at frequencies up to 100 MHz. |
| IEC 60754-1 | Halogen gas emission. |
| IEC 60754-2 | Smoke corrosivity. |
| IEC 60793-1-1 | Optical fibres — part 1: generic specification. |
| IEC 60794-1-1 | Optical cables — part 1: generic specification. |
| IEC 60874-1 | Connectors for optical fibres and cables — part 1: generic specification. |
| IEC 61024-1 | Protection of structures against lightning. |
| IEC 61034 | Smoke density and evolution. |
| IEC 61156-1 | Multicore and symmetrical pair/quad cables for digital communications — part 1: generic specification. |
| IEC 61280-4-1 | Fibre optic communication subsystem basic test procedure. |
| IEC 61312 | Protection against lightning electromagnetic impulse. |
| IEC 61935 | Generic specification for testing of generic cabling in accordance with ISO/IEC 11801 — part 1: test methods — part 2: patch cord and work area cables. |
| NEMA WC-63.1 | Performance standards for twisted pair premise voice and data communications cable. |
| NEMA WC-63.2 | Performance standards for coaxial communications cable. |
| NEMA WC-66.1 | Performance standards for category 6, category 7 $100\,\Omega$ shielded and unshielded twisted pair cables. |

| | |
|---|---|
| TIA/EIA 568-A | Commercial building telecommunications cabling standard, 1995. |
| TIA/EIA TSB 67 | Telecommunications systems bulletin. Transmission performance specifications for field testing of unshielded twisted pair cabling systems, October 1995. |
| TIA/EIA TSB 72 | Telecommunications systems bulletin. Centralized optical fiber cabling guidelines, October 1995. |
| TIA/EIA TSB 75 | Telecommunications systems bulletin. Additional horizontal cabling practices for open offices, August 1996. |
| TIA/EIA TSB 95 | Telecommunications systems bulletin. Additional transmission performance guidelines for $100\,\Omega$ 4 pair category 5 cabling, August 1999. |
| TIA/EIA 568 Addenda | |
| Addendum 1 | Propagation delay and delay skew, September 1997. |
| Addendum 2 | Corrections and additions to TIA-568-A, August 1998. |
| Addendum 3 | Hybrid cables, December 1998. |
| Addendum 4 | Patch cord qualification test, August 1999. |
| Addendum 5 | Additional transmission performance specifications for 4-pair $100\,\Omega$ category 5e cabling, November 1999. |
| TIA-IS 729 | Additional requirements for $100\,\Omega$ screened twisted pair cabling, March 1999. |
| TIA-569 | Commercial building standard for telecommunications pathways and spaces. |
| TIA-570 | Residential and light commercial telecommunications wiring standard. |
| TIA-606 | Administration standard for the telecommunications infrastructure of commercial buildings. |

| TIA-607 | Commercial building grounding/bonding requirements. |
| UL 910 | Test for flame propagation and smoke density values for electrical and optical fiber cables used in spaces transporting environmental air, i.e. Plenum. |
| UL 1581 | Reference standard for electrical wires, cables and flexible cords. |
| UL 1666 | Test for flame propagation height of electrical and optical fiber cables installed vertically in shafts. |

# Appendix II: Contact addresses for standards organisations and other interested bodies

| | |
|---|---|
| AFNOR | Association Française de Normalisation<br>Tour Europe<br>92049 Paris la Défense Cedex<br>France<br>Tel   +33 1 42 91 55 55<br>Fax  +33 1 42 91 56 56<br>www.afnor.fr |
| ANSI | American National Standards Institute<br>11 West 42nd Street<br>13th Floor<br>New York<br>NY 10036<br>USA<br>Tel   +1 212 642 4900<br>Fax  +1 212 302 1286<br>www.ansi.org |
| BICSI | BICSI<br>8610 Hidden River Parkway<br>Tampa<br>Fl 336337-1000<br>USA |

|  | Tel   +1 813 979 1991 |
|---|---|
|  | Fax   +1 813 971 4311 |
|  | www.bicsi.org |
| BSI | British Standards Institution |
|  | 389 Chiswick High Road |
|  | London W4 4AL |
|  | UK |
|  | Tel   +44 181 996 9000 |
|  | Fax   +44 181 996 7460 |
|  | www.bsi.org.uk |
| CEN | European Committee for Standardisation |
|  | Rue de Stassart 36 |
|  | B-1050 Brussels |
|  | Belgium |
|  | Tel   +32 2 550 08 11 |
|  | Fax   +32 2 550 08 19 |
|  | www.cenorm.be |
| CENELEC | CENELEC |
|  | Rue de Stassart 35 |
|  | B-1050 Brussels |
|  | Belgium |
|  | Tel   +32 2 519 6871 |
|  | Fax   +32 2 519 6919 |
|  | www.cenelec.be |
| CSA | Canadian Standards Association |
|  | 178 Rexdale Boulevard |
|  | Etobicoke |
|  | ON M9W 1R3 |
|  | Canada |
|  | Tel   +1 416 747 4044 |
|  | Fax   +1 416 747 2475 |
|  | www.cssinfo.com |
| Dansk Standard | Dansk Standard (DS) Electrotechnical Sector |
|  | Kollegievej 6 |
|  | DK-2920 Charlottenlund |
|  | Denmark |

Tel    +45 39 96 61 01
Fax    +45 39 96 61 02
www.ds.dk

DIN

Deutsches Institut für Normung e. V.
Burggrafenstrasse 6
10787 Berlin
Germany
www.din.de

EIA

Electronic Industries Alliance
2500 Wilson Boulevard
Arlington
VA 22202-3834
USA
Tel    +1 703 907 7500
Fax    +1 703 907 7501
www.eia.org

ETCI

Electro-Technical Council of Ireland
Unit 43
Parkwest Business Park
Dublin 12
Ireland
Tel    +353 1 623 99 01
Fax    +353 1 623 99 03
www.ecti.ie

ETSI

European Telecommunications Standards
Institute
Route de Lucioles
F-06921 Sophia Antipolis Cedex
France
Tel    +34 4 9294 4200
Fax    +34 4 9365 4716
www.etsi.fr

EUROPACABLE

The European Confederation of Associations of
Manufacturers of Insulated Wire and Cable
c/o CABLEBEL asbl
Diamant Building 5th Floor
Bld August Reyers 80

|       |                                          |
|-------|------------------------------------------|
|       | B-1030 Brussels                          |
|       | Belgium                                  |
|       | Tel    +32 2 702 62 25                   |
|       | Fax   +32 2 702 62 27                    |
| FCC   | Federal Communications Commission        |
|       | 1919 M Street NW                         |
|       | Room 702                                 |
|       | Washington                               |
|       | DC 20554                                 |
|       | USA                                      |
|       | Tel    +1 202 418 0200                   |
|       | Fax   +1 202 418 0232                    |
|       | www.fcc.gov                              |
| FIA   | Fibre Optic Industry Association         |
|       | Owles Hall                               |
|       | Owles Lane                               |
|       | Buntingford                              |
|       | Herts SG9 9PL                            |
|       | UK                                       |
|       | Tel    +44 1763 273039                   |
|       | Fax   +44 1763 273255                    |
|       | www.fibreoptic.org.uk                    |
| ICEA  | Insulated Cable Engineers Association    |
|       | PO Box 440                               |
|       | South Yarmouth                           |
|       | MA 0266                                  |
|       | USA                                      |
|       | Tel    +1 508 394 4424                   |
|       | Fax   +1 508 394 1194                    |
|       | icea@capecod.net                         |
| IEC   | International Electrotechnical Commission |
|       | Rue de Varembe, 3                        |
|       | PO Box 131                               |
|       | CH-1211 Geneva 20                        |
|       | Switzerland                              |
|       | Tel    +41 22 919 02 11                  |

|      | Fax   +41 22 919 03 00 |
|------|------------------------|
|      | www.iec.ch |
| IEEE | Institute of Electrical and Electronic Engineers |
|      | 445 Hoes Lane |
|      | PO Box 1331 |
|      | Piscataway |
|      | NJ 08855-1331 |
|      | USA |
|      | Tel   +1 732 981 0060 |
|      | Fax   +1 732 981 9667 |
|      | www.ieee.org |
| ISO  | International Organisation for Standardisation |
|      | Rue de Varembe, 1 |
|      | CH-1211 Geneva 20 |
|      | Switzerland |
|      | Tel   +41 22 749 01 11 |
|      | Fax   +41 22 733 34 30 |
|      | www.iso.ch |
| ITU  | International Telecommunications Union |
|      | Place des Nations |
|      | CH-1211 Geneva 20 |
|      | Switzerland |
|      | Tel   +41 22 730 51 51 |
|      | Fax   +41 22 733 72 56 |
|      | www.itu.ch |
| NEC  | Nederlands Elektrotechnisch Comite |
|      | Kalfjeslaan 2 |
|      | Postbus 5059 |
|      | NL — 2600 GB Delft |
|      | The Netherlands |
|      | Tel   +31 15 269 03 90 |
|      | Fax   +31 15 269 01 90 |
|      | www.nni.nl |
| NEK  | Norsk Elektroteknisk Komite |
|      | Harbitzalleen 2A |

|       | |
|-------|---|
|       | Postboks 280 Skoyen |
|       | N-0212 Oslo |
|       | Norway |
|       | Tel   +47 22 52 69 50 |
|       | Fax   +47 22 52 69 61 |
|       | www.nek.no |
| NEMA  | National Electrical Manufacturers Association |
|       | 1300 North 17th Street, Suite 1847 |
|       | Rosslyn |
|       | VA 22209 |
|       | USA |
|       | Tel   +1 703 841 3200 |
|       | Fax   +1 703 841 3300 |
|       | www.nema.org |
| NFPA  | National Fire Protection Agency |
|       | 1 Batterymarch Park |
|       | PO Box 9101 |
|       | Quincy |
|       | MA 02269-9101 |
|       | USA |
|       | Tel   +1 617 770 3000 |
|       | Fax   +1 617 770 0700 |
|       | www.nfpa.org |
| NNI   | Nederlands Normalisitie-instituut |
|       | Kalfjeslaan 2 |
|       | Postbus 5059 |
|       | NL — 2600 GB Delft |
|       | The Netherlands |
|       | Tel   +31 15 269 03 90 |
|       | Fax   +31 15 269 01 90 |
|       | www.nni.nl |
| NSF   | Norges Standardiseringsforbund |
|       | Drammensveien 145 |
|       | Postboks 353 Skoyen |
|       | N-0213 Oslo |
|       | Norway |

|                    |                                                |
|--------------------|------------------------------------------------|
|                    | Tel   +47 22 04 92 00                          |
|                    | Fax   +47 22 04 92 11                          |
|                    | www.standard.no                                |
| SEK                | Svenska Elektriska Kommissionen                |
|                    | Kistagangen 19                                 |
|                    | Box 1284                                       |
|                    | S-164 28 Kista Stockholm                       |
|                    | Sweden                                         |
|                    | Tel   +46 84 44 14 00                          |
|                    | Fax   +46 84 44 14 30                          |
|                    | www.sekom.se                                   |
| SESKO              | Finnish Electrotechnical Standards Association |
|                    | Sarkiniementie 3                               |
|                    | PO Box 134                                     |
|                    | SF-00211 Helsinki                              |
|                    | Finland                                        |
|                    | Tel   +358 9 696 391                           |
|                    | Fax   +358 9 677 059                           |
|                    | www.sesko.fi                                   |
| SFS                | Finnish Standards Association SFS              |
|                    | Maistraatinportti 2                            |
|                    | FIN-00240 Helsinki                             |
|                    | Finland                                        |
|                    | Tel   +358 9 149 9331                          |
|                    | Fax   +358 9 146 4925                          |
|                    | www.sfs.fi                                     |
| SIRIM              | SIRIM Berhad                                   |
|                    | 1 Persiaran Dato'Menteri                       |
|                    | PO Box 7035                                    |
|                    | Section 2                                      |
|                    | 40911 Shah Alam                                |
|                    | Malaysia                                       |
|                    | Tel   +60 3 559 2601                           |
|                    | Fax   +60 3 550 8095                           |
|                    | www.sirim.my                                   |
| Standards Australia | Standards Australia                           |
|                    | PO Box 1055                                    |

|       | Strathfield |
|-------|-------------|
|       | NSW 2135 |
|       | Australia |
|       | Tel    +61 2 9746 4700 |
|       | Fax   +61 2 9746 8540 |
|       | www.standards.com.au |
| SNZ   | Standards New Zealand |
|       | 155 The Terrace |
|       | Private Bag 2439 |
|       | Wellington |
|       | New Zealand |
|       | Tel    +64 4 498 5990 |
|       | Fax   +64 4 498 5994 |
|       | www.standards.co.nz |
| TIA   | Telecommunications Industry Association |
|       | 2500 Wilson Boulevard, Suite 315 |
|       | Arlington |
|       | VA 22201-3836 |
|       | USA |
|       | Tel    +1 703 907 7700 |
|       | Fax   +1 703 907 7727 |
|       | www.tiaonline.org |
| UL    | Underwriters Laboratories Inc. |
|       | 333 Pfingsten Road |
|       | Northbrook |
|       | IL 60062 |
|       | USA |
|       | Tel    +1 847 272 8800 |
|       | Fax   +1 847 272 8129 |
|       | www.ul.com |
| UTE   | Union Technique de l'Electricité |
|       | 33, Av. Général Leclerc — BP 23 |
|       | F-92262 Fontenay-aux-Roses Cedex |
|       | France |
|       | Tel    +33 1 40 93 62 00 |
|       | Fax   +33 1 40 93 44 08 |
|       | www.ute-fr.com |

VDE                    Deutsche Elektrotechnische Kommission im
                       DIN und VDE
                       Stresemannallee 15
                       D-60 596 Frankfurt am Main
                       Germany
                       Tel   +49 69 63 080
                       Fax   +49 69 63 12 925
                       www.dke.de

Obtaining copies of standards

The standards-writing bodies may be contacted directly or may
usually be obtained through the national standards body of the
country in which you reside. They are listed above under the following headings:

| | |
|---|---|
| Australia | Standards Australia |
| Canada | CSA |
| Denmark | Dansk Standard |
| Finland | SESKO, SFS |
| France | AFNOR, UTE |
| Germany | DIN, VDE |
| Holland | NEC, NNI |
| Ireland | ETCI |
| Malaysia | SIRIM |
| New Zealand | SNZ |
| Norway | NEK, NSF |
| Sweden | SEK |
| UK | BSI |

Alternatively, most standards can be purchased through:

Global Engineering Documents
Customer Support
15 Inverness Way
Englewood

CO 80112
USA
Tel +1 800 624 3974
Fax +1 303 792 2192
www.global.ihs.com

# Index

*Cable engineering for local area networks* provides a complete guide to the design, procurement, installation and testing procedures for local area networks (LANs) using both copper and optical fibre cable technology. International, European and American LAN and premises cabling standards are explained and compared, including the latest Category 5, Category 6 and Category 7 proposals. The latest standards in testing, electromagnetic compatibility (EMC) compliance and fire safety are also discussed in detail. By describing the theory as well as the practical issues involved, this book is an unrivalled source of information for those who need to understand the complexities of structured premises cabling systems used in today's office-based LANs. It will be essential reading for students, technicians, IT managers, systems designers and cable installation engineers.

There are chapters on the following subjects: basic mathematics; basic physics both electrical and optical; communications theory; local area networking and associated cabling; copper cable technology: cable, components, transmission and testing; optical cable technology: optical fibre, optical cable, components and testing; and cable system design and international standards. Appendices list relevant standards and the contact addresses of standards organisations worldwide.

Barry Elliott, a Chartered Electrical Engineer, has 20 years' experience in cabling and communications engineering and has worked for the Civil Aviation Authority and Ferranti Computer Systems. He is a BICSI Registered Communications Distribution Designer and holds an MBA from Henley Management College. Barry is currently Technical Marketing Manager at Brand-Rex Ltd.

Woodhead Publishing Ltd
Abington Hall
Abington
Cambridge
CB1 6AH
England
www.woodhead-publishing.com          ISBN 1 85573 488 5

WOODHEAD PUBLISHING LIMITED

Printed and bound by CPI Group (UK) Ltd, Croydon, CR0 4YY

03/10/2024

01040436-0016